Thermoelectric Polymers
Properties and Applications

Edited by

Inamuddin[1], Tariq Altalhi[2], Mohammad Abu Jafar Mazumder[3,4]

[1]Department of Applied Chemistry, Zakir Husain College of Engineering and Technology, Faculty of Engineering and Technology, Aligarh Muslim University, Aligarh-202002, India

[2]Department of Chemistry, College of Science, Taif University, 21944 Taif, Saudi Arabia

[3]Chemistry Department, King Fahd University of Petroleum & Minerals, Dhahran 31261, Saudi Arabia

[4]Interdisciplinary Research Center for Advanced Materials, King Fahd University of Petroleum & Minerals, Dhahran 31261, Saudi Arabia

Copyright © 2024 by the authors

Published by **Materials Research Forum LLC**
Millersville, PA 17551, USA

Published as part of the book series
Materials Research Foundations
Volume 162 (2024)
ISSN 2471-8890 (Print)
ISSN 2471-8904 (Online)

Print ISBN 978-1-64490-300-1
eBook ISBN 978-1-64490-301-8

Distributed worldwide by

Materials Research Forum LLC
105 Springdale Lane
Millersville, PA 17551
USA
https://www.mrforum.com

Manufactured in the United States of America
10 9 8 7 6 5 4 3 2 1

Table of Contents

Preface

Thermoelectric polymers are specifically designed hybrid materials capable of generating energy. For direct energy conversion between heat and electricity, thermoelectric materials can be utilized as active components in thermoelectric generators and as Peltier coolers. Aside from inorganic thermoelectric materials, thermoelectric polymers have gained considerable interest due to their unique benefits, such as low cost, high mechanical flexibility, lightweight, little or no toxicity, and inherently low heat conductivity. The power factor of thermoelectric polymers has steadily increased, and the highest ZT value at room temperature exceeds 0.25. The power factor can be further improved by forming composites with nanomaterials.

This book summarizes current advances in thermoelectric polymers and polymer composites. It also provides a review of recent developments in thermoelectric polymers and polymer composites. The book examines the link between thermoelectric characteristics and material structure, including chemical composition, microstructure, dopants, doping levels, methods of fabrication, thermoelectric effect, thermoelectric device conversion efficiency, and thermoelectric properties of conducting polymers. There are seven chapters in this book summarized as given below:

Chapter 1 deals with the thermoelectric properties of polymeric materials. Different thermoelectric effects, such as Seebeck, Peltier, Thomson effects, and others, are discussed at length. Additionally, a concise outline regarding various thermoelectric materials, including hybrid polymers, conductive polymers, and thermoelectric plastics, has also been elucidated.

Chapter 2 details various methods to obtain inorganic/organic composites or hybrid materials. The major focus is given to synthesis pathways such as electrospinning, solution processing, hydrothermal, hot pressing, atomic layer deposition, and three-dimensional printing techniques. Besides, the characterization of these hybrid composites based on mechanical, thermal, and microscopy properties is mentioned.

Chapter 3 describes thermoplastic polymers along with organic and inorganic composites. It also explains thermoelectric properties like the thermoelectric effect, Seebeck effect, Peltier effect, and Thomson effect. Moreover, the measurement techniques are also discussed in detail.

Chapter 4 summarizes thermoelectric polymers, their definition, preparatory methods, and processing conditions. It also states the p- and n-type thermoelectric polymers along with examples. Moreover, different parameters affecting the thermoelectric properties, such as polymer structure and concentration of polymer, are also discussed.

Chapter 5 is an initiative to explain the essence of caged compounds with their striking features and a plethora of applications.

Chapter 6 discusses the thermoelectric phenomenon, the figure of merit associated with the materials and devices, and the thermoelectric conversion efficiency. This chapter also focuses on various challenges in this domain along with their possible solutions.

Chapter 7 discusses how thermoelectric materials can be used for harvesting waste heat into electricity. The thermoelectric properties (electrical conductivity and Seebeck coefficient) and thermal properties of various promising n- and p-type polymeric compounds are discussed in detail. Moreover, emerging thermoelectric applications in different fields of life are also discussed.

Thermoelectric Polymers: Properties and Applications   Materials Research Forum LLC
Materials Research Foundations 162 (2024) 1-23   https://doi.org/10.21741/9781644903018-1

Chapter 1

Thermoelectric Effects

Monalisha Samanta, Subhajit Kundu, Debarati Mitra*

Department of Chemical Technology, University of Calcutta, 92 A P C Road, Kolkata-700009, West Bengal, India

debarati.che@gmail.com*

Abstract

The effect of thermoelectricity causes the conversion of waste heat into electricity. It is an economical environmentally beneficial and convenient energy conversion technology that can be used across a broad range of temperature. This chapter introduces thermoelectricity while also exploring its historical context. The thermoelectric property and the different thermoelectric effects, viz. Seebeck, Peltier, Thomson effects, and others are discussed. Additionally, a quick overview of thermoelectric materials has been presented. The figure of merit, a parameter defining research in thermoelectric effects is discussed in this chapter. After all, a concise outline regarding different thermoelectric materials including hybrid polymers, conductive polymers, thermoelectric plastics etc. has also been included herewith.

Keywords

Thermoelectric Effects, Energy Conversion, Performance Parameters, Hybrid Thermoelectric Materials, Thermoelectric Plastics

Contents

1.   Introduction

The interaction of electrical current as well as heat transfer in different conductors causes thermoelectric (TE) effects. The instantaneous transformation of heat to electricity is possible due to a coupling characteristic, termed thermopower; again, cooling is affected by applying an electric field across a thermoelectric material [1]. During the late 1950s, the development of semiconductors as thermoelectric sparked a resurgence of thermoelectrics. Goldsmid and Douglas as well as Ioffe et al., considered the concepts of both thermodynamics and solid-state, to extend earlier research on this arena, to the microscopic level, thereby paving the way for material science and its practical applications [2,3].

T.J. Seebeck identified the foremost thermoelectric phenomenon in 1821. He demonstrated that heating the interface between two distinct conductive materials might create an electromotive force [4]. After thirteen years of the discovery made by Seebeck, J. Peltier, a watchmaker of France, identified the second thermoelectric effect. He observed that based on the direction of an electric current in a thermocouple a minor heating as well as cooling effect can be created [5]. The interdependence of the Seebeck and Peltier phenomena however, was detected by W. Thomson in 1855. He applied thermodynamics concept and introduced a coefficient combining the effects of both Seebeck and Peltier. His hypothesis also revealed that in a homogeneous conductor, there is a third thermoelectric effect [6]. The Thomson effect is the combined effect of a temperature gradient in a current carrying conductor, which causes heating and cooling alternatively.

Altenkirch studied the process of energy conversion by thermocouples and reported that by enhancing the differential Seebeck coefficient's value, ramping up both the branches' electrical conductivity, and decreasing the thermal conductivities of both the branches.

The efficiency of a thermocouple can be enhanced [7,8]. Peltier refrigerators were only practicable around 1950s, after the use of semiconductors as thermoelectrics [9]. Investigation on thermocouples used as semiconductors also led to the development of high-efficiency thermoelectric generators for specialized applications. Despite this, thermoelectric energy converters' performance has always lagged below that of the greatest conventional machines [10].

The capacity of directly transfer heat energy into electrical energy of the thermoelectric materials, has increased their popularity. Waste heat can be converted to electric power using thermoelectric materials thereby improving energy efficiency [11]. Due to their low density, high flexibility and low toxicity, polymer-based thermo-electric materials especially for portable gadgets are particularly enticing. [12]. There are various advantages associated with polymer-based thermo-electric materials. Polymer derivatives, carbon nano-tube/conductive polymer composites, as well as inorganic semi-conductive nano-material/polymer composites are chosen among a wide variety of organic thermo-electric materials and grouped into four categories based on their thermoelectric performance, discussed in section 2.3[13, 14].

Conjugated polymers have semiconducting properties along with single and double loaded interphase, such as aromatic as well as compounds with heteroaromatic rings. On the other hand, polymers with side chains are easily soluble in organic solvents maintaining space in the polymeric chains after deposition, allowing them to control the degree of electronic interaction. [15]. Electrical conductivity is increased by the delocalization of Π electrons that are expected to occur because of conjugation in the conducting polymer. The degree of doping, the nature of the material used for doping, the structure, the density of the charge carriers, the temperature etc. all influence the material's conductivity. Thermoelectric performance is increased by the introduction of these polymers. As a result of doping, distinct electronic states appear in the polymer band gap [16, 17].

The conjugated lengths of the polymer chain can regulate the polymer's emission wavelength. It rises when the difference between HOMO and LUMO narrows. In the fabrication of organic lasers, this notion is quite beneficial [18, 19].

Thermoelectric plastics are also a type of polymer-based material that can change heat into electricity or the other way around. The thermo-electric gadgets made of inorganic materials are costly and include hazardous metals like lead, tellurium, and antimony, limiting their usage to high-end watches, arctic lights, and generators, among other things [20]. Organic materials, on the other hand, are tempting for application in thermoelectric devices because of their low heat conductivity, high flexibility, lightweight, non-toxicity, and ease of solution processing [21].

2.   Thermoelectric effects

The definitive transformation of temperature gradient into electrical voltage via a thermocouple and the reverse is known as thermoelectric effect [22]. A voltage is formed when a temperature differential is maintained across a material and while a temperature gradient is created between two ends of the substance [23].

The different kinds of thermoelectric effects are Seebeck, Peltier, and Thomson effects.

Seebeck Effect: Seebeck effect is considered to be the pillar of thermoelectric effect. This phenomenon was discovered in 1821 by Thomas Johann Seebeck (the German physicist). When a temperature differential exists between two junctions of two separate materials, Seebeck discovered that an electrical voltage is produced across the circuit's two junctions. [4,23,24]. The voltage created is directly proportional to the temperature difference between the junctions (Fig. 1). The constant of proportionality called Seebeck coefficient, is a characteristic property of the circuit. The expression of Seebeck effect is:

$$E = S(T_H - T_C) \tag{1}$$

where, E indicates the produced voltage between the two junctions of the circuit (volt, V); $(T_H - T_C)$ represents the temperature gradient (Kelvin, K) at the hot and cold junctions and S, the Seebeck coefficient (V/K).

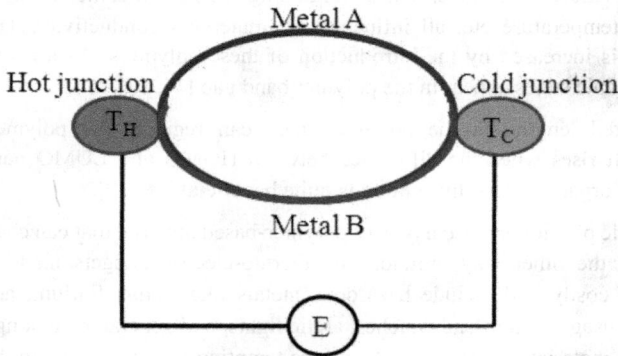

Figure 1: Schematic diagram of Seebeck effect

There are two types of Seebeck effect viz., non-isobaric and isobaric Seebeck effects [25]. In conducting membranes, two separate thermoelectric effects, viz., isobaric and non-isobaric, exist and a relationship between the coefficients of isobaric Seebeck and the non-isobaric Seebeck was established. The non-isobaric Seebeck coefficient has a greater range than the isobaric Seebeck coefficient which may be as high as, 10 mV/K. Large non-isobaric Seebeck coefficient values are observed in conducting nanoporous membranes (pore diameters from 5 to 200 nm) [25].

Using Seebeck effect, temperatures can be perceived as thermocouples by using one junction to detect while the other is maintained under isothermal condition. A thermopile can be made of many such thermocouples connected in series that have the ability to sense very small temperature variations [26,27].

Thermoelectric generators have some specific application in various fields:

- Waste heat is converted into extra electrical power in power plants

- Automotive thermoelectric generators are used to improve vehicle fuel economy

- In portable energy producing equipment for e.g., body heat-powered electronic devices [28,29].

Peltier Effect: Jean-Charles-Athanase Peltier, French physicist (1834) first noticed that the presence of a temperature gradient between two junctions of two dissimilar materials results in a flow of current through these materials (Fig. 2). The direction of current through the circuit determines the absorption or emission of heat at the junctions [5,23,24]. The rate at which heat is absorbed or emitted from the two junctions is as follows:

$$Q = \pi_{AB}I = (\pi_B - \pi_A)I \tag{2}$$

where, Q represents the quantity of heat generated per unit time (watt); I represents the electrical current (ampere) and π_{AB} is the Peltier coefficient (volt).

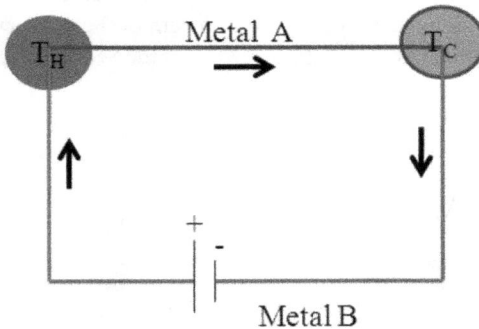

Figure 2: Schematic diagram of Peltier effect

The U.S. physicist Percy Williams Bridgman discovered a phenomenon - when an electric current passes through an anisotropic crystal, causing heat to be absorbed or released due to non-uniform current distribution; this is known as the Bridgeman effect or the internal Peltier effect [30].

One of the major applications of Peltier effect is in the construction of thermoelectric coolers. On either side, one n-type as well as one p-type semiconductor with varying electron densities are positioned parallel to one another and electrically coupled to form a series, installed on a thermally conducting plate [31]. When a voltage is supplied to the terminal ends of the two semiconductors, a direct current flows through the junction resulting in a temperature differential. At normal temperature, the side of the cooling plate is heated, this heat is eventually transported to the opposite side into the heat sink. More connections boost the system's cooling capacity. It is also used in temperature controller [32].

Thomson Effect: British physicist William Thomson discovered the Thomson effect in the year of 1851. Heat release or heat absorption takes place through a conductor carrying current having different temperatures along its length (Fig. 3) [23,24,33].

According to following equation, rate of heat emission and/or absorption is governed by the temperature differential as well as the current density via the conductor,

$$q = \rho J^2 - \mu J \frac{\Delta T}{\Delta x} \tag{3}$$

where, q is the rate of heat absorption or heat emission (joule/second); J is the current density (ampere/m^2); ρ symbolizes the resistance of the substance (ohm); $\Delta T/\Delta x$ stands for the temperature differential throughout the length of the conductor; μ represents the Thomson coefficient that has identical units (V/K) as the Seebeck coefficient.

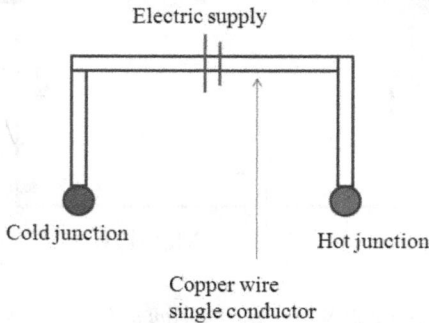

Figure 3: Schematic diagram of Thomson effect

Hence, the Seebeck, Peltier, and Thomson coefficients are related as:

$$\pi_{AB} = ST \tag{4}$$

$$\mu = T\frac{dS}{dT} \tag{5}$$

The combination of these three thermoelectric effects give rise to a complete thermoelectric phenomenon. These effects are exhibited by all metals and also some polymeric thermoelectric materials [23].

2.1   Performance parameters of thermoelectric material

Figure of merit (zT): Ioffe first discovered (1949) a factor, the dimensionless figure of merit (zT) for determining the performance of thermoelectric material [23,24,34]. The figure of merit (zT) is expressed as:

$$zT = \frac{S^2\sigma}{k}T \tag{6}$$

here, $S^2\sigma$ represents thermopower also called the power factor; S (V/K) represents the Seebeck coefficient; σ(Siemens/meter or S/m) stands for the electrical conductivity; T stands for the absolute temperature (K); k represents the thermal conductivity (W/m-K) of the substance. Thermoelectric materials with high Seebeck coefficient and high electrical conductivity, but low thermal conductivity, are ideal since they have a high zT value (k).

Seebeck Coefficient (S): The Seebeck coefficient (V/K) is a voltage generated per unit temperature gradient, applied throughout a substance. A high value of Seebeck coefficient, (~150–250 V/K or above), is required for a perfect thermoelectric [23,34].

Electrical conductivity (σ): Electrical conductivity (S/m) is a major material-dependent characteristic that determines the capacity of a material to conduct electricity [23,24]. The electrical conductivity is expressed as:

$$\sigma = e(\mu_e n + \mu_h p) \tag{7}$$

where e (Coulomb or C) is the electronic charge, and μ_e(m^2/V·s) is the electron mobility; μ_h (m^2/V·s) represents the hole mobility; n is the electron density and p is the hole density. The electrical conductivity of a suitable thermoelectric material is typically about 10^5 S/cm.

Thermal conductivity (k): The capacity of a substance to conduct heat is determined by its thermal conductivity. The total contribution from charge carriers such as electrons (k_c) or holes, as well as phonons (k_l), determines the overall thermal conductivity of a conductor/semiconductor [23,24],

or,

$$k = k_c + k_1 \tag{8}$$

The values of thermal conductivity that might render the conductors suitable for use as thermoelectric materials are typically k<2W m^{-1}K^{-1} as well as k_l should be nearly equal to k_c. According to the Wiedmann-Franz law, the charge thermal conductivity (k_c) is correlated with electrical conductivity as:

$$k_c = L\sigma T \tag{9}$$

here, L represents the Lorentz constant (2.45×10^{-8}V^{2}K^{-2} as well as 1.5×10^{-8}V^{2}K^{-2} for metals and non-degenerate semiconductors respectively). Increased electrical conductivity results in enhanced thermal conductivity, as a result of charge-carriers [35].

2.2　Thermoelectric materials

The most common thermoelectric material is bismuth telluride (Bi$_2$Te$_3$). Bismuth telluride was discovered in the year 1954 by Goldsmid [2]. Till date, bismuth telluride-based thermoelectric materials are of great importance in this field [3,23].

Ioffe et al. (1956) suggested that an alloy of a semiconductor and an isomorphic material can possess high thermoelectric performance due to low thermal conductivity without altering the electrical conductivity [36]. Several semiconductor alloys, can be used as thermoelectric materials over a broad range of temperature. The most commonly used alloying semiconductor materials are p-type e.g. Bi$_{2-x}$Sb$_x$Te$_3$ and n-type e.g. Bi$_2$Te$_{3-x}$Se$_x$ having zT values ~1 at 300 K and these are the best materials for thermoelectric applications [37-41].

Other thermoelectrics include lead telluride (PbTe) based materials; they are also called lead chalcogenides like PbSe, PbS, and their alloys. These materials are operable at high temperature (600-800K) [40,41]. Doping of thermoelectric materials has resulted in improvement of their properties. Thallium (Tl) doped PbTe based thermoelectrics have high zT value (1.5) at 773K [42]. When $PbTe_{0.85}Se_{0.15}$ was doped with sodium, a zT value of 1.8 at 850K was reached [43]. Since lead is toxic, it is being replaced by GeTe-rich alloys viz. TAGS (GeTe-$AgSbTe_2$) which have zT value ~1 around the same temperature range [44].

The elemental Si and Ge exhibit low thermoelectric performance because these have high thermal conductivity. However, the alloy SiGe offers good thermoelectric performance, with high thermal conductivity and remarkably high thermal stability, and a melting point of 1027^0C. Of late, n-type nanostructured phosphorous is doped with SiGe alloy to form SiGe bulk alloy to enhance the value of zT (to ~1.3) under a high temperature of 1173K [45].

Other Si based materials include Mg_2Si as well as its solid solutions, which have larger manganese silicides (HMS). They are useful for thermoelectric applications, have low cost and are environmentally friendly. The general formula of Mg_2Si and its solid solutions is $(Mg,Ca)_2(Si,Ge,Sn)$ [46-48].

Some crystalline as well as amorphous solids (glass) commonly known as phonon glass electron crystal (PGEC) material are expected to serve as excellent thermoelectrics [49]. The PGEC thermoelectric materials have complex structures like skutterudites, clathrates etc., with big voids in the structures of their crystal. Clathrates type materials have low thermal conductivity, with cage crystal lattice of foreign atoms. The guest atoms (Ga, Al, Sn, Si or Ge), serve as phonon scattering sites thereby reducing their thermal conductivity [50-53]. Skutterudites are symbolized as TX_3; T stands for transition metals like Fe, Co etc., X represents a metalloid which belongs to Periodic Table group 15 (P, Sb or As). The zT value for a skutterudite like material ($Ba_{0.08}Yb_{0.04}La_{0.05}Co_4Sb_{12}$) is 1.8 at higher temperature (400-850K) [54-57].

Some organic materials like PANI (polyaniline), PEDOT poly (3, 4-ethylenedioxythiophene) are useful in thermoelectric applications. Liu and co-researchers (2012) introduced a new concept - 'Phonon Liquid Electron Crystal' (PLEC) that explains the good thermoelectric performance (zT is 1.5 at 1000K) of Cu_2Se; hence binary Cu-based chalcogenides, i.e. $Cu_{2-\delta}X$ where X represents Se, Se and Te are widely used in thermoelectric applications [58-63].

2.3   Hybrid thermoelectric materials

More recently, the demand for hybrid thermoelectric materials has increased particularly in the power/energy sector, to meet the requirements of the increasing population of the world. The hybrid thermoelectric materials are formed mainly from different types of polymeric materials [23].

The hybrid polymeric thermoelectric (TE) materials have several benefits over their metallic or ceramic counterparts. Polymeric materials, for example, have substantially lower densities and heat conductivities than metals or ceramics. They are also flexible. For thermoelectric energy conversion applications, different types of conducting polymers have been used like polyaniline (PANI), poly(3,4-ethylenedioxythiophene) (PEDOT), polythiophene (PTh), polycarbazole (PC), polypyrrole (PPY), polyacetylene (PA) and their composites [25,64].

Hybrid composite materials can be made by combining conducting polymers with inorganic nanoparticles. For example, spin-coating chemically produced PANI with 10–50 wt% CuO nanoparticles resulted in hybrid composite thin films. In thermopower measurement, PANI exhibits p-type behavior whereas the CuO nanoparticles display the n-type [65]. If the polymers are not conducting material, then the electrical conductivity of composite material or polymer can be improved by adding a metal based thermoelectric component. For example, a polysiloxane polymer matrix composite film filled with micrometer-scale thermoelectric $Sn_{0.85}Sb_{0.15}O_2$ particles. The percolation action of the metallic filler causes the electrical conductivity of the insulating polymer to rise by several orders of magnitude [66]. A hybrid composite of thermoelectric thick film (inorganic material) and conducting polymer (organic material) can be fabricated using the screen-printing technique [67].

To create a hybrid composite material, a conducting polymer called poly(3,4-ethylenedioxythiophene)-poly(styrenesulfonate) (PEDOT-PSS) can be inserted into the micropores of the annealed thick thermoelectric films that are screen-printed [25]. Aranguren and his co-researchers fabricated a flexible thermoelectric energy generator which was based on the π-shaped PEDOT thermoelectric module [68]. For improving the electrical conductivity (σ) and the Seebeck coefficient, sometimes graphite is added to polymers [69]. To enhance the thermoelectric performance of PEDOT-PSS, a reagent, choline chloride (ChCl)-urea mixture was inserted by surface treatment [70].

One of the most promising thermoelectric polymers is PANI since it has high thermoelectricity, good electrical conductivity, highly stable in air, easy fabrication, good redox reversibility with controllable properties and low cost [71-76]. Generally, PANI has been explored for the purpose of making capacitors, battery electrodes, chemical sensors, electrochromic devices, enzyme immobilizers, light-emitting diodes and anticorrosion reagents [77-83]. However, PANI has limited applications due to its unsolved decoupling issue that alters its, electrical conductivity, thermal conductivity as well as Seebeck coefficient [25].

To alleviate these challenges, it has recently been proposed that organic and inorganic materials be combined with PANI to form nanocomposites. Some examples are Bi_2Te_3/PANI and Sb_2S_3/PANI that have been effectively produced by electrochemical reactions bearing enhanced electrical conductivity [84,85].

Normally, when fossil fuel is burned, waste heat energy is released at somewhat more than 150°C. At moderate temperatures, organic, polymeric and hybrid thermoelectric materials are capable of extracting useful electricity from waste heat, as against the conventional thermoelectric materials that generally operate at high temperature. There are different types of hybrid thermoelectric materials that can be utilized to produce electricity from waste heat [23]:

- Hybrid of conducting polymers with metal nanoparticles
- Hybrid of conducting polymers with Bi_2Te_3 nanoparticles
- Hybrid of conducting polymers with carbon nano-tubes (CNTs)
- Hybrid of other inorganic-organic molecules

Hybrid of conducting polymers with metal nanoparticles: Hybrid thermoelectric materials with increased electrical conductivity and thermal stability are made by combining metal nanoparticles with different polymers [86-91]. Moreover, between the conducting polymeric chains, these metallic nanoparticles operate as charge carriers. To generate hybrid films, polyaniline doped in camphor sulfonic acid was mixed with a solution of poly-N-vinyl pyrrolidone (PVP) in Pt or Au nanoparticles in m-cresol [92]. Since PVP has an insulating effect, a small amount of PVP should be added, otherwise the electrical conductivity of these films will be reduced drastically. Hence, PANI is generally used to protect the reagent in preference to PVP [93,94].

Other polymers that might be used in a metal nanoparticle hybrid material include, poly (3,4-ethylenedioxythiphene)-poly and styrene sulfonate (PEDOT-PSS). Here, nanoparticles (PEDOT-PSS) can be dissolved in water. For these thermoelectrics, mercaptopropionic acid (MPA) as well as mercaptohexanoic acid serve as shielding agents and improve the electronic conductivity as compared to the virgin material [95,96]. Furthermore, the size of the nanoparticle has a considerable impact on the Seebeck coefficient as well as the electrical conductivity. Since the electrons of nanoparticles of Au may mix with the voids of PEDOT-PSS, the concentration of PEDOT-PSS film as carrier is reduced, thereby resulting in an increase in the Seebeck coefficient [97].

Hybrid of conducting polymers with Bi$_2$Te$_3$ nanoparticles: By embedding inorganic nanomaterials like Bi$_2$Te$_3$ in conducting polymers, the zT value of inorganic semiconductors (Bi$_2$Te$_3$) may be fundamentally changed, since these nano materials can considerably modify the density of states (DOS) of these inorganic nanomaterials [98]. The resultant hybrid materials take into account the benefits of conducting polymers and inorganic semiconductors. When an identical sample is made with the bulk Bi$_2$Te$_3$ in place of its nanoparticles, the zT is drastically reduced. However, the preparation of Bi$_2$Te$_3$ nanoparticles is a cumbersome process; while it has a high zT, the sample's value is dependent on how it is processed. Furthermore, it is unstable in presence of air. The value of zT is improved significantly when p- as well as n-type Bi$_2$Te$_3$ nanoparticles are hybridised with PEDOT-PSS. This forms a stable hybrid thermoelectric material which can operate at ambient temperature [2].

Hybrid of conducting polymers with carbon nano-tubes (CNTs): Carbon nano-tubes have high value of electrical conductivity as well as thermal conductivity when used as a single tube, but exhibit poor conductivity when utilized as a bucky paper sheet [98-104]. Hybrids of polymers with carbon nano-tubes have gained importance to achieve enhanced thermoelectric performance because of these fascinating features. To increase the thermoelectric characteristics, carbon nano-tube hybrids with polymers like poly(3-hexylthiophene) (P3HT), PEDOT-PSS and PANI are investigated. Phonon scattering occurs at the carbon nano-tube-polymer interface in all three cases, resulting in a considerable drop in thermal conductivity and a high value of zT [23].

Hybrids of other organic/inorganic molecules: The most attractive materials with outstanding thermoelectric features are combinations of organic compounds with n-type inorganic semiconductors. For example, layered TiS$_2$ can be combined with certain organic materials. Next, externally introducing electrons into TiS$_2$ makes it n-doped. Organic cations, such as hexylammonium, stabilise the hybrid material by acting as n-type carriers during current as well as energy transmission [105,106].
All of the materials mentioned above are in the research and development stage.

2.4   Thermoelectric plastics

Thermoelectric plastics are polymer-based materials that can convert heat to electricity or the other way around. Thermoelectric devices made of inorganic materials are costly and include dangerous metals like lead, tellurium, and antimony, limiting their usage to high-end watches that harvest body heat, arctic lights, and generators, among other things.

Though organic materials or polymers have low thermal conductivity but high flexibility, nontoxic nature and lightweight; render them suitable for making of thermoelectric devices. Waste heat from different systems is converted into electrical energy using thermoelectric materials. They are made up of abundant components such as sulfur, nitrogen, oxygen and carbon. Automobiles loose a huge amount of energy; therefore we require materials possessing high thermoelectric efficiency and low CO_2 emissions to achieve global sustainability. Thermoelectric plastics are used in a variety of systems that help to retrieve lost heat in different industries capable to a range of ~200°C, such as chimneys, power lines and cladding of pipes. Organic semiconductors, particularly conductive fillers, conjugated polymers, insulating polymers and dopant counterion are the basic building components of thermoelectric plastics. Mostly, thermoelectric plastics are π-conjugated polymers with smaller molecules such as PEDOT, PANI, PPY, PAC, polythiophene, pentacene, fullerene, etc. [2]. However, the poor energy conversion efficiency of polymer-based materials, continue to limit their thermoelectric applications.

Conclusion

The thermoelectric effect is an aspect in which an electric potential is created by a temperature differential or an electric current causing a temperature difference. The Seebeck effect (generating a voltage from a temperature differential), Peltier effect (producing heat flow with an electric current), and Thomson effect are the names for these processes (existence of both electric current and a temperature difference in a system causing alternate heating and cooling in a conducting material). Low-cost materials with a strong thermoelectric effect have been explored for applications in power generation and refrigeration. Bismuth telluride (Bi_2Te_3) is the most commonly used thermoelectric material. Some polymeric thermoelectric materials are being explored due to their high zT, Seebeck coefficient and electrical conductivity values and low thermal conductivity. Conducting polymers have been suggested as a way to improve the thermoelectric effect. Inorganic semi-conductive nano-materials/polymer composites, PEDOT as well as carbon nano-tube/conductive polymer composites are the examples of polymer-based thermoelectric materials. Furthermore, novel nano-fillers with a greater Seebeck coefficient or electrical conductivity have the ability to enhance the power factor of organic composites even further. Flexible thermoelectric materials offer a wide range of applications, including power generators that capture thermal energy and power electronics, because to their improved energy conversion efficiency and simpler process ability.

Reference

[1] D. Dai, Y. Zhou, J. Liu, Liquid metal based thermoelectric generation system for waste heat recovery, Renew. Energy. 36 (2011) 3530-3536. https://doi.org/10.1016/j.renene.2011.06.012

[2] H.J. Goldsmid, R.W. Douglas, The use of semiconductors in thermoelectric refrigeration. Br. J. Appl. Phys. 5 (1954) 386-390. https://doi.org/10.1088/0508-3443/5/11/303

[3] A.F. Ioffe, Semiconductor Thermoelements and Thermoelectric Cooling, Infosearch, Ltd., London, 1957.

[4] T.J. Seebeck, The magnetic polarization of metals and ores produced by temperature difference, Proc. Prussian Acad. Sci. (1822) 265-373.

[5] J.C.A. Peltier, Nouvelles experiences sur la caloricite des courants electrique, Ann. Chim. Phys. 56 (1834) 371-386.

[6] W. Thomson, On the mechanical theory of thermo-electric currents, Math. Phys. Pap. 1 (1851) 316-323.

[7] A. Fledhoff, Power conversion and its efficiency in thermoelectric materials, Entropy. 22 (2020) 803. https://doi.org/10.3390/e22080803

[8] K. Landecker, Some aspects of the performance of refrigerating thermojunctions with radial flow of current, J. Appl. Phys. 47 (1976) 1846. https://doi.org/10.1063/1.322903

[9] M. Gaikwad, D. Shevade, A. Kadam, B. Shubham, Review on thermoelectric refrigeration: Materials and technology, Int. J. Curr. Eng. Technol, 2016.

[10] G. Tan, M. Ohta, M.G. Kanatzidis, Thermoelectric power generation: From new materials to devices, Philos. Trans. Royal Soc. A Philos T R Soc A., 2019. https://doi.org/10.1098/rsta.2018.0450

[11] S. LeBlanc, Thermoelectric generators: Linking material properties and systems engineering for waste heat recovery applications, Sustain. Mater. Technol. 1 (2014) 26-35. https://doi.org/10.1016/j.susmat.2014.11.002

[12] J. Li, A.B. Huckleby, M. Zhang, Polymer-based thermoelectric materials: A review of power factor improving strategies, J. Materiom. 8 (2022) 204-220. https://doi.org/10.1016/j.jmat.2021.03.013

[13] C.J. Yao, H.L. Zhang, Q. Zhang, Recent progress in thermoelectric materials based on conjugated polymers, Polymers. 11 (2019) 107. https://doi.org/10.3390/polym11010107

[14] M. Goel, M. Thelakkat, Polymer thermoelectrics: Opportunities and challenges, Macromolecules. 53 (2020) 3632-3642. https://doi.org/10.1021/acs.macromol.9b02453

[15] B. Russ, A. Glaudell, J.J. Urban, M.L. Chabinyc, R.A. Segalman, Organic thermoelectric materials for energy harvesting and temperature control, Nat. Rev. Mater. 1 (2016) 16050. https://doi.org/10.1038/natrevmats.2016.50

[16] P. Sengodu, A.D. Deshmukh, Conducting polymers and their inorganic composites for advanced Li-ion batteries: A review, RSC Adv. 5 (2015) 42109-42130. https://doi.org/10.1039/C4RA17254J

[17] S. Wang, G. Zuo, J. Kim, H. Sirringhaus, Progress of conjugated polymers as emerging thermoelectric materials, Prog. Polym. Sci. 129 (2022) 101548. https://doi.org/10.1016/j.progpolymsci.2022.101548

[18] W. Zhao, J. Ding, Y. Zou, C.-a. Di, D. Zhu, Chemical doping of organic semiconductors for thermoelectric applications, Chem. Soc. Rev. 49 (2020) 7210-7228. https://doi.org/10.1039/D0CS00204F

[19] X.L. Shi, W.Y. Chen, T. Zhang, J. Zou, Z.-G.Chen, Fiber-based thermoelectrics for solid, portable, and wearable electronics, Energy Environ. Sci. 14 (2021) 729. https://doi.org/10.1039/D0EE03520C

[20] S.K. Tripathi, R. Kaur, Organic Semiconductors and polymers, in: R. Kumar, R. Singh (Eds.), Thermoelectricity and Advanced Thermoelectric Materials, Woodhead Publishing, Elsevier, 2021, 195-196. https://doi.org/10.1016/B978-0-12-819984-8.00002-3

[21] H. Wang, C. Yu, Organic thermoelectric materials and devices, in: R. Funahashi (Eds.), Thermoelectric Energy Conversion, Woodhead Publishing Series in Electronic and Optical Materials, Elsevier, 2021, 347-365. https://doi.org/10.1016/B978-0-12-818535-3.00005-0

[22] Thermoelectric effect, https://www.newworldencyclopedia.org/entry/Thermoelectric_effect/, 2020 (accessed on 04th May, 2022).

[23] R. Kumar, R. Singh, Thermoelectricity and Advanced Thermoelectric Materials, Woodhead Publishing, Elsevier, United Kingdom, 2021.

[24] L. Chen, R. Liu, X. Shi, Thermoelectric Materials and Devices, Elsevier, United Kingdom, 2021. https://doi.org/10.1016/B978-0-12-818413-4.00006-5

[25] Y.X. Gan, Nanomaterials For Thermoelectric Devices, Pan Stanford Publishing Pte. Ltd., Taylor & Francis, New York, 2018. https://doi.org/10.1201/9780429488726

[26] D. Camuffo, Measuring Temperature, Microclimate for Cultural Heritage, Elsevier, 2014. https://doi.org/10.1016/B978-0-444-63296-8.00012-3

[27] B. J. Huang, A precise measurement of temperature difference using thermopiles, Exp. Therm. Fluid. Sci. 3 (1990) 265-271. https://doi.org/10.1016/0894-1777(90)90001-N

[28] D. Enescu, Thermoelectric Energy Harvesting: Basic Principles and Applications, Green Energy Advances, IntechOpen, 2019. https://doi.org/10.5772/intechopen.83495. https://doi.org/10.5772/intechopen.83495

[29] M. Hyland, H. Hunter, J. Liu, E. Veety, D. Vashaee, Wearable thermoelectric generators for human body heat harvesting, Appl. Energy. 182 (2016) 518-524. https://doi.org/10.1016/j.apenergy.2016.08.150

[30] Bridgman Effect, https://tabroot.com/bridgman-effect/, 2022 (accessed on 10th May, 2022).

[31] O. Y. Titov, Y. G. Gurevich, Temperature gradient and transport of heat and charge in a semiconductor structure, Low Temp. Phys. 47 (2021) 550. https://doi.org/10.1063/10.0005182

[32] S. Ahmad, M.E.S. Abdullah, M.F. Yaakub, A.Z. Jidin, S.H. Joharin, M. Zahari, Analysis of portable temperature-controlled device by using peltier effect, Proc. Mechanical Eng. Res. Day, 2017, 176-177.

[33] E. Lenz, Einige Versuche Im Gebiete Des Galvanismus, Ann. Phys. 120 (1838) 342-349. https://doi.org/10.1002/andp.18381200612

[34] E. M. Barber, Thermoelectric Materials Advances and Applications, CRC Press, Taylor & Francis, New York, 2015.

[35] G.S. Nolas, H.J. Goldsmid, Thermal Conductivity: Theory, Properties and Applications, Springer, New York, 2004.

[36] U. Birkholz, Untersuchung der intermetallischen Verbindung Bi2Te3 sowie der festen Losungen Bi2-xSbxTe3 und Bi2Te3-xSex hinsichtlich ihrer Eignung als Material fur Halbleiter-Thermoelemente, Z. Naturforsch. 13 (1958) 780-792. https://doi.org/10.1515/zna-1958-0910

[37] F.D. Rosi, B. Abeles, R.V. Jensen, Materials for thermoelectric refrigeration, J. Phys.Chem. Solids. 10 (1959) 191. https://doi.org/10.1016/0022-3697(59)90074-5

[38] C. Han, Q. Sun, Z. Li, S.X. Dou, Thermoelectric enhancement of different kinds of metal chalcogenides, Adv. Energy Mater. 6 (2016) 1600498. https://doi.org/10.1002/aenm.201600498

[39] H.S. Kim, N.A. Heinz, Z.M. Gibbs, Y. Tang, S.D. Kang, G.J. Snyder, High thermoelectric performance in (Bi0.25Sb0.75)2Te3 due to band convergence and improved by carrier concentration control, Mater. Today. 20 (2017) 452-459. https://doi.org/10.1016/j.mattod.2017.02.007

[40] A.D. LaLonde, Y. Pei, H. Wang, G. Jeffrey Snyder, Lead telluride alloy thermoelectrics, Mater. Today. 14 (2011) 526-532. https://doi.org/10.1016/S1369-7021(11)70278-4

[41] Y. Pei, H. Wang, G.J. Snyder, Band engineering of thermoelectric materials, Adv. Mater. 24 (2012) 6125-6135. https://doi.org/10.1002/adma.201202919

[42] J.P. Heremans, V. Jovovic, E.S. Toberrer, A. Saramat, K. Kurosaki, A. Charoenphakdee, S. Yamanaka, G.J. Snyder, Enhancement of thermoelectric efficiency in PbTe by distortion of the electronic density of states, Science. 321 (2008) 554-557. https://doi.org/10.1126/science.1159725

[43] Y. Pei, X. Shi, A.D. LaLonde, H. Wang, L. Chen, G.J. Snyder, Convergence of electronic bands for high performance bulk thermoelectrics, Nature. 473 (2011) 66-69. https://doi.org/10.1038/nature09996

[44] E.A. Skrabek, D.S. Trimmer, in: D.M. Rowe (Eds.), CRC Handbook of Thermoelectrics, CRC Press, Boca Raton, 1994.

[45] C.B. Vining, Thermoelectric materials-silicon germanium, in: D.M. Rowe (Eds.), Handbook of Thermoelectric, CRC Press LLC, Danvers, 1995. https://doi.org/10.1201/9781420049718.ch28

[46] W. Liu, K. Yin, Q. Zhang, C. Uher, X. Tang, Eco-friendly high-performance silicide thermoelectric materials, Natl. Sci. Rev. 4 (2017) 611-626. https://doi.org/10.1093/nsr/nwx011

[47] T. Itoh, M. Yamada, Synthesis of thermoelectric manganese silicide by mechanical alloying and pulse discharge sintering, J. Electron. Mater. 38 (2009) 925-929. https://doi.org/10.1007/s11664-009-0697-3

[48] W. Luo, H. Li, Y. Yan, Z. Lin, X. Tang, Q. Zhang, C. Uher, Rapid synthesis of high thermoelectric performance higher manganese silicide with in-situ formed nano-phase of MnSi, Intermetallics. 19 (2011) 404-408. https://doi.org/10.1016/j.intermet.2010.11.008

[49] G.S. Nolas, J. Sharp, H.J. Goldsmid, The phonon-glass electron-crystal approach to thermoelectric materials research, in: G.S. Nolas , J. Sharp, H.J. Goldsmid (Eds.), Thermoelectrics: Basic Principles and New Materials Developments, Springer, New York, 2001, pp. 177-207. https://doi.org/10.1007/978-3-662-04569-5_6

[50] H. Kleinke, New bulk materials for thermoelectric power generation: Clathrates and complex antimonides, Chem. Mater. 22 (2010) 604-611. https://doi.org/10.1021/cm901591d

[51] E. Toberer, M. Christensen, B.B. Iversen, G.J. Snyder, High temperature thermoelectric efficiency in Ba8Ga16Ge30, Phys. Rev. B. 77 (2008) 075203.

[52] G.S. Nolas, G.A. Slack, S.B. Schujman, T.M. Tritt (Eds.), Recent Trends in Thermoelectric Materials Research I-Semiconductor and Semimetals, Academic Press, London, 2001.

[53] B.X. Shi, J. Yang, S. Bai, J. Yang, H. Wang, M. Chi, W. Zhang, L. Chen, W. Wong-Ng, On the design of high-efficiency thermoelectric clathrates through a systematic cross-substitution of framework elements, Adv. Funct. Mater. 20 (2010) 755-763. https://doi.org/10.1002/adfm.200901817

[54] J. Graff, S. Zhu, T. Holgate, J. Peng, J. He, T.M. Tritt, High-temperature thermoelectric properties of Co4Sb12-based skutterudites with multiple filler atoms: Ce0.1InxYbyCo4Sb12, J. Electron. Mater. 40 (2011) 696-701. https://doi.org/10.1007/s11664-011-1514-3

[55] G. Tan, L.-D. Zhao, M.G. Kanatzidis, Rationally designing high-performance bulk thermoelectric materials, Chem. Rev. 116 (2016) 12123-12149. https://doi.org/10.1021/acs.chemrev.6b00255

[56] J.L. Mi, T.J. Zhu, X.B. Zhao, J. Ma, Nanostructuring and thermoelectric properties of bulk skutterudite compound CoSb3, J. Appl. Phys. 101 (2007) 054314. https://doi.org/10.1063/1.2436927

[57] J. Yang, X. Shi, S. Bai, W. Zhang, L. Chen, US patent 0071741 A1, 2010 March 25.

[58] M. Scholdt, H. Do, J. Lang, A. Gall, A. Colsmann, U. Lemmer, Organic semiconductors for thermoelectric applications, J. Electron. Mater. 39 (2010) 1589-1592. https://doi.org/10.1007/s11664-010-1271-8

[59] L. Jun, L.M. Zhang, L. He, X.F. Tang, Synthesis and thermoelectric properties of polyaniline, J. Wuhan Univ. Technol. Mater. Sci. Ed. 18 (2003) 53-55. https://doi.org/10.1007/BF02838459

[60] H. Liu, X. Shi, F. Xu, L. Zhang, W. Zhang, L. Chen, Q. Li, C. Uher, T. Day, G.J. Snyder, Copper ion liquid-like thermoelectrics, Nat. Mater. 11 (2012) 422-425. https://doi.org/10.1038/nmat3273

[61] B. Yu, W. Liu, S. Chen, H. Wang, H. Wang, G. Chen, H. Ren, Thermoelectric properties of copper selenide with ordered selenium layer and disordered copper layer, Nano Energy. 1 (2012) 472-478. https://doi.org/10.1016/j.nanoen.2012.02.010

[62] Y. He, T. Day, T.S. Zhang, H. Liu, X. Shi, L. Chen, G.J. Snyder, High thermoelectric performance in non-toxic earth-abundant copper sulfide, Adv. Mater. 26 (2014) 3974-3978. https://doi.org/10.1002/adma.201400515

[63] Y. He, T. Zhang, X. Shi, S.H. Wei, L.D. Chen, High thermoelectric performance in copper telluride, NPG Asia Mater. 7 (2015) 210. https://doi.org/10.1038/am.2015.91

[64] C. Gao, G. Chen, Conducting polymer/carbon particle thermoelectric composites: Emerging green energy materials, Compos. Sci. Technol. 124 (2016) 52-70. https://doi.org/10.1016/j.compscitech.2016.01.014

[65] D.M. Jundale, S.T. Navale, G.D. Khuspe, D.S. Dalavi, P.S. Patil, V.B. Patil, Polyaniline-CuO hybrid nanocomposites: synthesis, structural, morphological, optical and electrical transport studies, J. Mater. Sci.: Mater. Electron. 24 (2013) 3526-3535. https://doi.org/10.1007/s10854-013-1280-5

[66] B. Plochmann, S. Lang, R. Ruger, R. Moos, Optimization of thermoelectric properties of metal-oxide-based polymer composites, J. Appl. Polym. Sci. 131 (2014) 40038. https://doi.org/10.1002/app.40038

[67] J.H. We, S.J. Kim, B.J. Cho, Hybrid composite of screen printed inorganic thermoelectric film and organic conducting polymer for flexible thermoelectric power generator, Energy. 73 (2014) 506-512. https://doi.org/10.1016/j.energy.2014.06.047

[68] P. Aranguren, A. Roch, L. Stepien, M. Abt, M. von Lukowicz, I. Dani, D. Astrain, Optimized design for flexible polymer thermoelectric generators, Appl. Therm. Eng. 102 (2016) 402-411. https://doi.org/10.1016/j.applthermaleng.2016.03.037

[69] L. Wang, D. Wang, G. Zhu, J. Li, F. Pan, Thermoelectric properties of conducting polyaniline/graphite composites, Mater. Lett. 65 (2011) 1086-1088. https://doi.org/10.1016/j.matlet.2011.01.014

[70] Z. Zhu, C. Liu, Q. Jiang, H. Shi, J. Xu, F. Jiang, J. Xiong, E. Liu, Green DES mixture as a surface treatment recipe for improving the thermoelectric properties of PEDOT:PSS films, Synth. Met. 209 (2015) 313-318. https://doi.org/10.1016/j.synthmet.2015.08.006

[71] L. Su, Y. X. Gan, Experimental study on synthesizing TiO2 nanotube/polyaniline (PANI) nanocomposites and their thermoelectric and photosensitive property characterization, Composites Part B: Eng. 43 (2012) 170-182. https://doi.org/10.1016/j.compositesb.2011.07.015

[72] C. H. Park, S. K. Jang, F. S. Kim, Conductivity enhancement of surface-polymerized polyaniline films via control of processing conditions, Appl. Surf. Sci. 429 (2018) 121-127. https://doi.org/10.1016/j.apsusc.2017.09.031

[73] H. Okamoto, T. Kotaka, Structure and properties of polyaniline films prepared via electrochemical polymerization: Effect of pH in electrochemical polymerization media on the primary structure and acid dissociation constant of product polyaniline films, Polymer. 39 (1998) 4349-4358. https://doi.org/10.1016/S0032-3861(98)00013-5

[74] S.L. Bai, Y.L. Tian, M. Cui, J.H. Sun, Y. Tian, R.X. Luo, A.F. Chen, D.Q. Li, Polyaniline@SnO2 heterojunction loading on flexible PET thin film for detection of NH3 at room temperature, Sens. Actuators B: Chem. 226 (2016) 540-547. https://doi.org/10.1016/j.snb.2015.12.007

[75] S. Mu, J. Kan, J. Lu, L. Zhuang, Interconversion of polarons and bipolarons of polyaniline during the electrochemical polymerization of aniline, J. Electroanal. Chem. 446 (1998) 107-112. https://doi.org/10.1016/S0022-0728(97)00529-9

[76] J. Zhang, L. Kong, B. Wang, Y. Luo, L. Kang, In-situ electrochemical polymerization of multi-walled carbon nanotube/ polyaniline composite films for electrochemical supercapacitors, Synth. Met. 159 (2009) 260-266. https://doi.org/10.1016/j.synthmet.2008.09.018

[77] W. Qiu, R. Zhou, L. Yang, Q. Liu, Lithium-ion rechargeable battery with petroleum coke anode and polyaniline cathode, Solid State Ionics. 86 (1998) 903-906. https://doi.org/10.1016/0167-2738(96)00211-1

[78] X. Zhang, L. Ji, S. Zhang, W. Yang, Synthesis of a novel polyaniline-intercalated layered manganese oxide nanocomposite as electrode material for electrochemical capacitor, J. Power Sources. 173 (2007) 1017-1023. https://doi.org/10.1016/j.jpowsour.2007.08.083

[79] M. Joubert, M. Bouhadid, D. Begue, P. Iratcabal, N. Redon, J. Desbrieres, S. Reynaud, Conducting polyaniline composite: from syntheses in waterborne systems to chemical sensor devices, Polymer. 51 (2010) 1716-1722. https://doi.org/10.1016/j.polymer.2010.01.052

[80] J. Jang, J. Ha, K. Kim, Organic light-emitting diode with polyaniline-poly(styrene sulfonate) as a hole injection layer, Thin Solid Films. 516 (2008) 3152-3156. https://doi.org/10.1016/j.tsf.2007.08.088

[81] X. Yang, B. Li, H. Wang, B. Hou, Anticorrosion performance of polyaniline nanostructures on mild steel, Prog. Org. Coat. 69 (2010) 267-271. https://doi.org/10.1016/j.porgcoat.2010.06.004

[82] L. Zhao, L. Zhao, Y. Xu, T. Qiu, L. Zhi, G. Shi, Polyaniline electrochromic devices with transparent graphene electrodes, Electrochim. Acta, 55 (2009) 491-497. https://doi.org/10.1016/j.electacta.2009.08.063

[83] F. N. Crespilho, R. M. Lost, S. A. Travain, Jr. O. N. Oliverira, V. Zucolotto, Enzyme immobilization on Ag nanoparticles/polyaniline nanocomposites, Biosens. Bioelectron. 24 (2009) 3073-3077. https://doi.org/10.1016/j.bios.2009.03.026

[84] K. Chatterjee, A. Suresh, S. Ganguly, K. Kargupta, D. Banerjee, Synthesis and characterization of an electro-deposited polyaniline bismuth telluride nanocomposite: A novel thermoelectric material, Mater. Charact. 60 (2009) 1597-1601. https://doi.org/10.1016/j.matchar.2009.09.012

[85] S. Subramanian, P.C. Lekha, D. P. Pddiyan, Enhanced electrical response in Sb2S3 thin films by the inclusion of polyaniline during electrodeposition, Physica B, 405 (2010) 925-931. https://doi.org/10.1016/j.physb.2009.10.016

[86] H. Hirai, Y. Nakao, N. Toshima, Preparation of colloidal rhodium in poly(vinyl alcohol) by reduction with methanol, J. Macromol. Sci. Chem. A.12 (1978) 1117-1141. https://doi.org/10.1080/00222337808063179

[87] N. Toshima, M. Harada, Y. Yamazaki, K. Asakura, Catalytic activity and structural analysis of polymer-protected Au-Pd bimetallic clusters prepared by the simultaneous reduction of HAuCl4 and PdCl2, J. Phys. Chem. 96 (1992) 9927-9933. https://doi.org/10.1021/j100203a064

[88] N. Toshima, T. Yonezawa, Bimetallic nanoparticles-novel materials for chemical and physical applications, New J. Chem. 22 (1998) 1179-1201. https://doi.org/10.1039/a805753b

[89] Y. Shiraishi, N. Toshima, Oxidation of ethylene catalyzed by colloidal dispersions of poly(sodium acrylate)-protected silver nanoclusters, Colloids Surf. A Physicochem. Eng. Asp. 169 (1-3) (2000) 59-66. https://doi.org/10.1016/S0927-7757(00)00417-9

[90] H. Zhang, T. Watanabe, M. Okumura, M. Haruta, N. Toshima, Catalytically highly active top gold atom on palladium nanocluster, Nat. Mater. 11 (2012) 49-52. https://doi.org/10.1038/nmat3143

[91] B. Corain, G. Schmid, N. Toshima (Eds.), Metal Nanoclusters in Catalysis and Materials Science: The Issue of Size-Control, Elsevier, Amsterdam, 2011.

[92] N. Toshima, N. Jiravanichanun, H. Marutani, Organic thermoelectric materials composed of conducting polymers and metal nanoparticles, J. Electron. Mater. 41 (2012) 1735-1742. https://doi.org/10.1007/s11664-012-2041-6

[93] T. Yonezawa, T.T. Kunitake, Practical preparation of anionic mercapto ligand-stabilized gold nanoparticles and their immobilization, Colloids Surf. A Physicochem. Eng. Asp. 149 (1999) 193-199. https://doi.org/10.1016/S0927-7757(98)00309-4

[94] N. Toshima, N. Jiravanichnun, Improvement of thermoelectric properties of PEDOT/PSS films with addition of gold nanoparticles: Enhancement of Seebeck coefficient, J. Electron. Mater. 42 (2013) 1882-1887. https://doi.org/10.1007/s11664-012-2458-y

[95] A. Yoshida, N. Toshima, Gold nanoparticle and gold nano rod embedded PEDOT:PSS thin films as organic thermoelectric materials, J. Electron. Mater. 43 (2014) 1492-1497. https://doi.org/10.1007/s11664-013-2745-2

[96] A. Yoshida, N. Toshima, Thermoelectric properties of hybrid thin films of PEDOT-PSS and silver nanowires, J. Electron. Mater. 45 (2016) 2914-2919. https://doi.org/10.1007/s11664-015-4326-z

[97] S. Ichikawa, N. Toshima, Improvement of thermoelectric properties of composite films of PEDOT-PSS with xylitol by means of stretching and solvent treatment, Polym. J. 47 (2015) 522-526. https://doi.org/10.1038/pj.2015.28

[98] M.S. Dresselhaus, G. Chen, M.Y. Tang, R.G. Yang, H. Lee, D.Z. Wang, Z.F. Ren, J.P. Fleurial, P. Gogna, New directions for low-dimensional thermoelectric materials, Adv. Mater. 19 (2007) 1043-1053. https://doi.org/10.1002/adma.200600527

[99] M. Prato, Fullerene chemistry for materials science applications, J. Mater. Chem. 7 (1997) 1097-1109. https://doi.org/10.1039/a700080d

[100] M.I. Katsnelson, Graphene: Carbon in two dimensions, Mater. Today. 10 (2007) 20-27. https://doi.org/10.1016/S1369-7021(06)71788-6

[101] K. Balasubramanian, M. Burghard, Chemically functionalized carbon nanotubes, Small. (2005) 180-192. https://doi.org/10.1002/smll.200400118

[102] S. Iijima, Helical microtubules of graphitic carbon, Nature. 354 (1991) 56-58. https://doi.org/10.1038/354056a0

[103] S. Hata, T. Omura, K. Oshima, Y. Du, Y. Shiraishi, N. Toshima, Novel preparation of poly (3, 4-ethylenedioxythiophene)-poly (styrene sulfonate)-protected noble metal nanoparticles as organic-inorganic hybrid thermoelectric materials, Bull. Soc. Photogr. Imaging. 27 (2017) 13-18.

[104] Y. Nakai, K. Honda, K. Yanagi, H. Kataura, T. Kato, T. Yamamoto, Y. Maniwa, Giant Seebeck coefficient in semiconducting single-wall carbon nanotube film, Appl. Phys. Exp. 7 (2014) 025103. https://doi.org/10.7567/APEX.7.025103

[105] C. Wan, X. Gu, F. Dang, T. Itoh, Y. Wang, H. Sasaki, M. Kondo, K. Koga, K. Yabuki, G.J. Snyder, R. Yang, K. Koumoto, Flexible n-type thermoelectric materials by organic intercalation of layered transition metal dichalcogenide TiS2, Nat. Mater. 14 (2015) 622-627. https://doi.org/10.1038/nmat4251

[106] T. Fukumaru, T. Fujigaya, N. Nakashima, Development of n-type cobaltocene encapsulated carbon nanotubes with remarkable thermoelectric property, Sci. Rep. 5 (2015) 7951. https://doi.org/10.1038/srep07951

Chapter 2

Fabrication of Polymer and Organic-Inorganic Composites

Moises Bustamante-Torres[1,a,*], Y. Aylin Esquivel-Lozano[1,b], Jorge Perea-Armijos[2,c], and Emilio Bucio[1,d,*]

[1] Department of Radiation Chemistry and Radiochemistry, Institute of Nuclear Science, National Autonomous University of Mexico, Mexico City 04510, Mexico

[2] School of Earth Sciences, Energy and Environment, Yachay Tech University, Urcuqui city, Ecuador

[a]moisesbustamante819@gmail.com, [b]aylinesquivel95@gmail.com, [c]jorge.perea.armijos@gmail.com, [d]ebucio@nucleares.unam.mx

Abstract

Polymers are macromolecules that can be classified in many ways. There are natural or synthetic and organic or inorganic materials among these ways. These materials can form a cross-linking forming a network capable of incorporating reinforcement materials. Combining these two materials is known as a composite or hybrid material, which represents a significant advance in materials science due to the possibility of improving the properties of an organic matrix by incorporating an inorganic filler material into it. This chapter details the main characteristics of natural, synthetic, organic, and inorganic polymers. Besides, the synthesis pathways to synthesize organic-inorganic composites such as electrospinning, solution processing, hydrothermal, hot pressing, atomic layer deposition, and three-dimensional printing techniques are described. Finally, the characterization of these hybrid composites based on mechanical, thermal, and microscopy are mentioned.

Keywords

Polymer, Composites, Organic-Inorganic Composites, Synthesis, Characterization

Contents

Thermoelectric Polymers: Properties and Applications   Materials Research Forum LLC
Materials Research Foundations 162 (2024) 24-55   https://doi.org/10.21741/9781644903018-2

1.   Introduction

Polymer science is one of the fastest-growing technology and most impactful fields because of its special applications [1]. Since the last century, composite materials have arisen as remarkable materials [2], thanks to the interaction of two or more distinct materials with a finite interface [3], reinforcement, and polymeric matrix [4]. These polymeric composites usually present interesting physicochemical and biological characteristics [5,6] for a wide range of applications.

Organic/inorganic composites also known as hybrid materials usually consist of organic polymer composites with inorganic nanoscale building blocks [7]. These hybrid materials have been studied because of their excellent mechanical strength and high thermal stability [8]. The polymer composites display significant friction and wear performance after modified with functional fillers and reinforcements [9].

The composites' processing and property characteristics are controlled by the interaction of polymer matrix (toughness and adhesion) and reinforcing phase [10, 11]. Besides, these organic matrices can incorporate filling particles (< 100 nm) [12]. Thermosets and thermoplastics are the main material class used a polymeric matrix [13]. Thermosetting materials present high mechanical, thermal, and stability properties due to their permanent network structure, while thermoplastic materials can often be readily melted and processed [14].

Inorganic-organic composite material incorporates filling particles (*e.g.* rigidity and thermal stability) into a polymer matrix (*e.g.* flexibility) [7, 15]. Several techniques have been carried out to obtain composite with desirable properties by combining different fillers and matrices. This chapter describes some of the main techniques applied to form organic and inorganic composites. Besides, the mechanical, thermal and microscopy characterization techniques of polymer composites are detailed.

2. Polymers

Polymers are defined as a long-chain backbone of monomers and side groups [16] with a remarkable versatility. Nonetheless, the polymer properties will depend on the chemical linkage type, chain length, and the nature of the end groups [17]. Polymers can be combined widely with many other polymers to improve synergistically their properties, as well as to provide a cost reduction and improvement in material processing, etc. [18]. Moreover, polymers can be combined with a vast amount of other compounds. In the last years, various applications in the biomedical area [19], wastewater treatment [20], agriculture [21], among others, have been applied using polymers, depending on the properties for each field.

Polymers have interesting properties such as high strength, toughness, resistance to corrosion, lack of conductivity [22]. Actually, mechanical properties are strongly linked to the molecular weight [23], and cross-linking degree. For example, highly cross-linked materials have a greater density and mechanical modulus [24]. Besides, those polymers with higher molecular weight will present better mechanical properties than the polymers with lower molecular weight [25].

Interesting thermal properties are present in polymers as a response of temperature changes. When the polymer is heated, the polymer chains can entangle, and the polymer becomes soft and pliable, similar to rubber. All polymers will transition from a glassy to a rubbery state at some temperature [26], called glass transition temperature (T_g). The T_g of a polymer increases with the degree of cross-linking [27]. Melting temperature (T_m) is a specific temperature that determines the solid material changes from a solid state to a liquid or melts [28]. It affects the efficiency of degassing through hydrogen solubility, diffusivity, and melt viscosity [29].

2.1 Organic polymers

Organic polymers are macromolecules containing partially carbon atoms into the primary backbone [30]. The use of organic polymers has many advantages, including an increased settlement rate, decreased costs, and environment properties [31]. They are primarily sorted out as natural biodegradable and synthetic nontoxic biodegradable polymers [32]. In fact, covalent organic polymers are categorized as a class of highly porous cross-linked materials [33]. Organic polymers are degraded into volatile combustible products when heated above certain critical temperatures, depending on their chemical structures [34]. These polymers are inexpensive, which has attracted the interest of scientists. Most organic polymers have conduct and semiconducting properties, making them valuable materials [35].

2.2 Inorganic polymers

Inorganic polymers, like organic polymers, comprise a chain of repeating units; however, they do not contain carbon within their backbone [36]. Inorganic polymers are outstandingly versatile materials that comprises silicon, phosphorus, and nitrogen into their structures [32, 37]. Inorganic polymers present remarkable properties such as good low-temperature flexibility, high thermal and oxidative stability [38]. Some examples of inorganic polymers are polysilanes (Si–Si bonds), polysiloxanes (Si–O bonds, or silicones), polysilazanes (Si–N bonds), polysulfides (S–S bonds), polyphosphazenes (P–N bonds) [30]. Figure 1 illustrates the chemical structure of some inorganic polymers.

Polysilane

Polysiloxanes

Polysilazanes

Polysulfides

Polyphosphazene

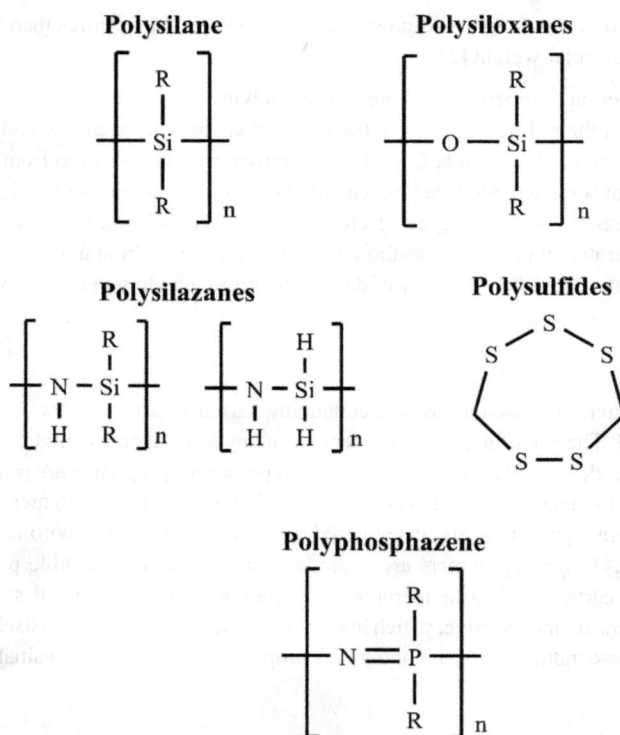

Figure 1. Chemical structure of some inorganic polymers.

2.2. Polymeric matrix

Polymeric matrix (PM) is a continuous phase cross-linked of polymers. The cross-linking of macromolecules is a critical factor, transforming the polymers into networks that can display mechanical properties [39] called PM. PM is composed of organic molecules classified as thermoset or thermoplastic [40]. In composites, the main function of PM is to retain and maintain the reinforcing agents [41], offering an exciting platform for the controlled delivery of drugs for sustainable formulation [42].

2.2.1 Thermoplastic

Thermoplastics polymers can be adjusted into new shapes due to the application of heat and can be processed either in the heat-softened state or in the liquid form [43]. Besides, they can also be recycled. However, thermoplastic polymers can lose their physical properties due to the breakage of polymeric chains [44]. These polymeric materials are characterized by high ductility and high impact strength, but poor creep resistance, which is a significant disadvantage. Figure 2a illustrate a cross-linking chain in thermoplastic compared to thermosetting.

In addition, thermoplastic polymers can be either crystalline, semicrystalline or amorphous. Crystallinity is a remarkable characteristic of thermoplastic polymers [45]. Crystalline thermoplastics are composed of molecular chains arranged in regular, organized regions connected by regions of disordered amorphous chains [46]. Changes in the crystallinity of thermoplastics can lead to significant changes in the mechanical properties of composites containing them [47]. The most common crystalline thermoplastic polymer is polypropylene, which is employed with glass fiber in conventional composites [48]. Nonetheless, for amorphous polymers the T_g is a reversible transition from a hard and brittle state into a molten or rubber state [49].

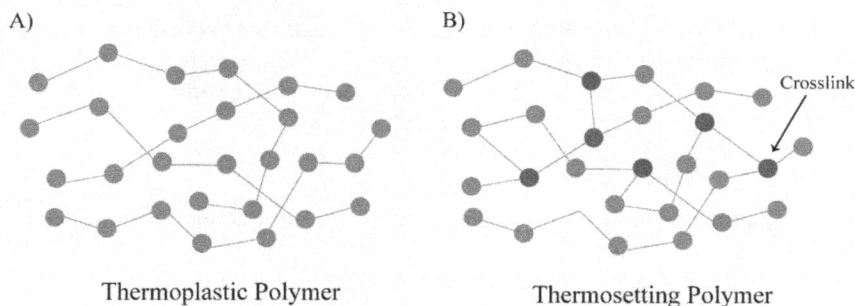

A)   B)   Crosslink

Thermoplastic Polymer   Thermosetting Polymer

Figure 2. Polymer chains in A) thermoplastic polymers, and B) thermosetting polymer with cross-linking chains.

2.2.2 Thermoset

Thermosets are three-dimensional cross-linking polymers promoted by chemical bonding between macromolecular chains [50]. Therefore, thermosets tend to be more resistant to solvents and corrosive environments than thermoplastics [13]. Figure 2b shows the internal cross-linking of thermosetting polymers, resulting into a cross-linking structure. However, the cross-linked structure of the polymer limits the movement of molecules, leading to low toughness and brittle materials [51, 52]. Some of the main thermoset materials are polyesters, epoxies, polyurethanes, and silicones [53].

3. Composite

A composite material consists of combining at least two materials [54]. These polymeric materials achieve the most interesting properties from the reinforcement and the matrix [55]. The opportunity to select the constituents make the optimum use of their properties as an attractive situation for multiple applications [56]. This process requires an extensive investigation of the physical and mechanical properties of each constituent [57]. Hydrogels have been used as a cross-linked polymeric matrix to incorporate inorganic nanoparticles into their degradable structure by improving their mechanical strength and incrementing high thermal stability [58].

3.1 Filled composites

Fillers are particulate materials added to polymers to reduce cost [59] and improve mechanical and physical properties. Impact and tensile strengths are usually decreased while hardness and stiffness increase [60], because the fillers are added at very high levels [61]. Besides fillers prevent convection and modify the rheology of the unreacted formulation [62]. When the filler is incompatible with the polymer, voids tend to occur at the interface, free volume system and high permeability [63]. Dispersed fillers represent less than 50% in a composite, modifying the matrix properties [64]. They could be oriented directly to fill in some of the void or surface of the matrices [65].

3.2 Reinforced composites

Reinforced composites have an inherent high strength and stiffness [66]. The possibility of fiber-reinforced composite adaptation is much easier than other materials [67]. Fibrous reinforcements are added to polymers to increase stiffness and strength [59]. These fibers may be random, oriented, or in a mat format [68]. However, glass and carbon fibers have limited end-of-life options, which is a significant concern for their environmental impact [69]. These components are not just superficial fibers and polymer matrix combinations; as synergistic effects are essential to their mechanical properties. In the case of continuous reinforcement, the fibers are mainly arranged on the surface layer in a unidirectional or woven fabric, resulting in excellent mechanical properties in the direction of the layer surface [70].

4. Organic-Inorganic composites

Organic-inorganic composites are composed of a mixture of inorganic components, organic components, or both types of elements [71]. These composites have been predominantly in the form of thin films grown on rigid substrates [72, 73, 74]. The human tooth is a hybrid inorganic/organic composite characterized by a complex hierarchical structure responsible [75, 76]. Homogeneously dispersed organic-inorganic hybrid composites have been synthetized by mixing different polymers, or by the appropriate choice of inorganic precursors [77]. Polymer-inorganic hybrids combine the advantages of inorganic and polymeric materials to be used as membrane materials [39]. Figure 3, illustrates a hybrid organic-inorganic composite.

Organic-Inorganic Hybrid composite

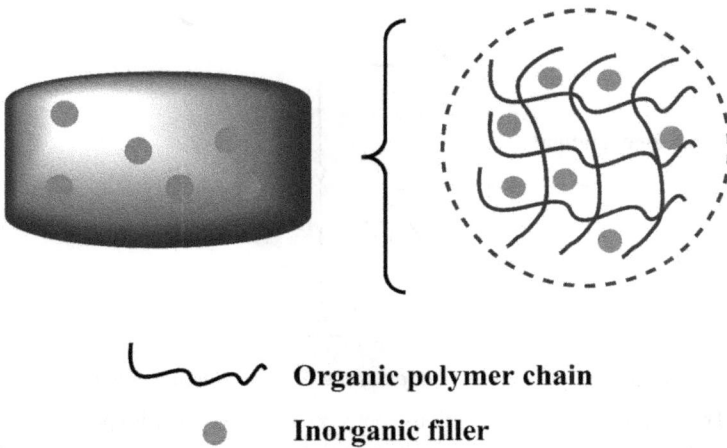

~~~ **Organic polymer chain**

● **Inorganic filler**

Figure 3. Schematic representations of an organic polymeric matrix containing inorganic filler.

4.1 Synthesis of inorganic-organic composites

4.1.1 Electrospinning technique

Electrospinning is a highly employed method to synthesize organic-inorganic hybrid composite [78]. Inorganic fibers can be prepared by electrospinning with polymers incorporated in the precursor solution [79]. This method allows obtaining fibers of the desired polymer at various sizes ranging from micrometers to nanometers [80]. Besides, it is influenced by the viscosity, the operating voltage, the temperature, the pressure, and the flow rate [81]. Besides, the distance between the collector and the syringe also influences the fibers' size [82].

Powerful energy sources, a syringe, and a disk or metal collector are required to carry out the electrospinning technique. The polymer solution is placed into the syringe, where the cathode is usually connected. Subsequently, the collector is placed far away from the syringe, connecting the anode to it. Finally, a voltage from the power sources is supplied, and the polymers are deposited in the collector [82]. Figure 4 illustrates the electrospinning process.

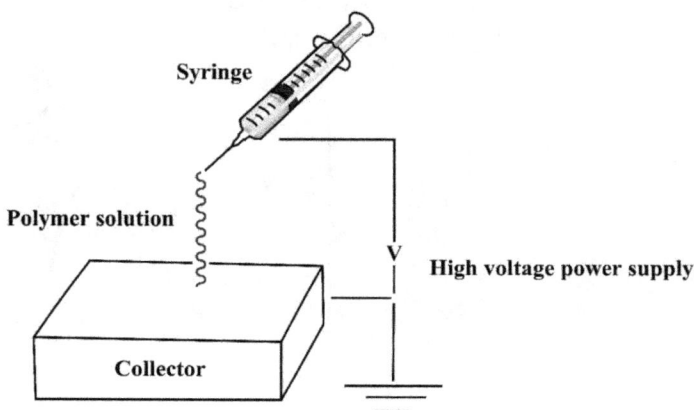

Figure 4. The schematization of the Electrospinning technique

Uslu and co-workers synthesized various epoxy resin composites with polymer nanofibers such as nylon 66 (N66), Polyacrylonitrile (PAN), polyvinyl alcohol (PVA), and polyvinylchloride (PVC), obtained through the electrospinning technique. Composite containing CN66 presented the best mechanical tests [83]. Cui et al. developed a nanofiber membrane composite made of PVA (polyvinyl alcohol) with LS (sodium lignosulfonate). This biodegradable membrane was evaluated to filter the air, resulting in 99.94% effectiveness [84]. On the other hand, PVA was reported as a scaffold for bone tissue regeneration. Calcium phosphate compound ($CaHPO_4 \cdot 2H_2O$) particles were incorporated into the PVA matrix, obtaining suitable biocompatibility and an appropriate place for dental pulp cells to grow [85].

The incorporation of inorganic agents into composites has been extensively studied since metallic agents provide a possible improvement in the properties of the matrix and add new ones, such as an antimicrobial action [86]. For instance, Zhang et al., reported a composite membrane made up of polylactic acid (PLA) and zinc oxide (ZnO) nanoparticles for applications in food packaging. These membranes obtained by electrospinning presented adequate mechanical properties and antimicrobial activity against *Escherichia coli* and *Staphylococcus aureus* [87].

4.1.2 Solution processing

Solution processing is a simple and low-cost technique to synthesize composites. This process has been to synthesize thin films and carry out coatings more efficiently [88]. During this process, the polymer or prepolymer must be soluble and the nanofillers swellable [89]. When the polymer and layered silicate are mixed, the polymer chains intercalate, displacing the solvent within the interlayer of the silicate [90]. Upon evaporating the solvent, polymer chains may reassemble, wrapping the nanofillers [91]. The homogeneous dispersion is achieved due to a suitable compatibility of solvent with polymer and filler plays [92]. Some other techniques are based on the solution principle to form composites such as the hydrothermal method, spray coating, and inkjet printing.

4.1.2.1   Hydrothermal synthesis

Hydrothermal process is produced due to chemical reactions in aqueous solution above the boiling point of water [93]. This process is carried out at a sealed pressure vessel at high temperature and pressure [94]. This technique is employed to produce inorganic and organic materials or composites. Chao and co-workers developed a molybdenum disulfide (MoS_2)/[poly(3,4-ethylene dioxythiophene): poly(styrenesulfonate (PEDOT: PSS) nanosheet composite to apply it as a supercapacitor [95].

4.1.2.2   Spray coating

This method consists of depositing the working solutions into a machine containing a nozzle. Aerosols will be produced and then sprayed into the required surface, forming the coating. Graphene composite films containing PEDOT: PSS were developed using the spray coating technique. They also stabilized the spray by adding isopropyl alcohol to obtain more homogeneous films, excellent conductive properties, and highly transparent [96]. One variation of spray coating is electrospray deposition (ESD). EDS system requires liquid atomization using electrical forces that generate a droplet plume by charging the liquid at a high voltage [97]. EDS uses tiny droplets, and to some extent, the charge, size, and motion of these droplets is controlled by electrical means [98]. EDS can control the film's thickness, uniformity, and morphology by adjusting the solution concentration and applied voltage [99].

4.2.3 Inkjet printing

Inkjet printing is an economical technology for fabricating multiple materials. This method is based on the scientific principle of ink: formation of ink drops, jet: the fall of that ink, and printing: droplet controlled into a land in the right place [100]. Materials can vary from a simple polymer solution to advanced nanoparticle dispersions [101]. This process depends directly on the type of media, ink set, printing speed, pre-and post-treatment of the media, and output method [102, 103].

Inkjet printing is a unique system that allows delivering a small amount of materials as ink to a specific substrate location [104]. Figure 5 illustrates the inkjet printing system. It is a non-contact method, reason by which it avoids a cross-contamination from the surface allowing a 3D printing [105, 106].

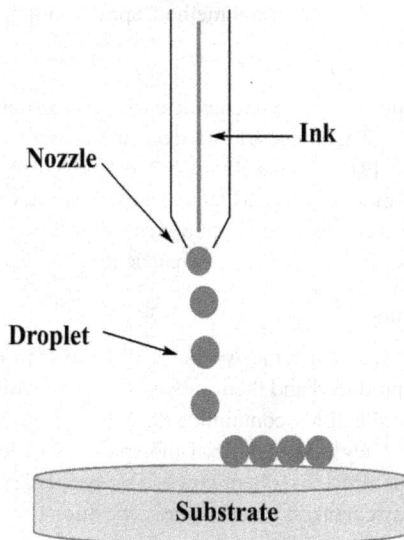

Figure 5. Inkjet printing representation.

Some works of hybrid composites containing organic-inorganic composites have been reported. For instance, Kato *et al.* prepared a coatable ink of PEDOT: PSS and an organic additive (polyacrylic acid) (PAA) [107]. Mikolajek et al. reported the synthesis of three composites ink containing $Ba_{0.6}Sr_{0.4}TiO_3$ (different concentrations 33, 50, 66 vol%) as the ceramic component and poly (methyl methacrylate) (PMMA) as the organic polymer through fully inkjet printing. These composites allow printing on flexible substrates [108].

Klein et al. synthesized sol-gel hybrid resin composites containing organic-inorganic phases for digital light processing (DLP) and inkjet printing. The inkjet printing method was used to improve the curved surfaces; therefore, the transmittance increased. Besides, several quantum dots were synthesized and incorporated into the resins [109].

4.1.4 Hot pressing

Hot pressing is a prominent technique for producing inorganic-organic composites. It employs an uniaxial application of pressure on a compact placed in a heated die to improve the densification [110]. Figure 6 represents the hot pressing techniques. During hot pressing, temperature and pressure are applied simultaneously to the powder compact in a graphite die [111]. To avoid the die damage, the graphite die is enclosed in a protective atmosphere [112]. To fabricate composites, the time consumption of pressing steps is 3–4 h, including cooling and heating [113]. By applying the pressure hydraulically, the points of contact between the materials increase, allowing the composite to present a high homogeneity. So, Song and co-workers developed a ceramic composite containing two different additives. Titanium diboride (TB_2) matrix was reinforced with Silicon carbide (SiC) and boron nitride (Bn) through the hot pressing technique at 2000 °C and 50 MPa. Due to the interaction between the TiB_2 and SiC phases and the high temperatures reached, SiO_2 was generated. The TiB_2-SiC composite presented a high hardness, while the $TiB2$-BN displayed a higher toughness [114].

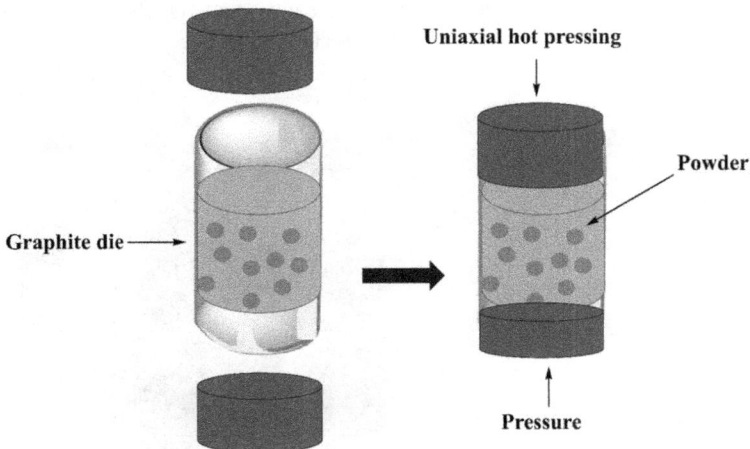

Uniaxial hot pressing

Powder

Graphite die

Pressure

Figure 6. Schematic the hot pressing technique in a graphite container with a hydraulic press.

Many materials can be synthesized through this process. Besides, the different parameters can be regulated to obtain composites with different characteristics. For example, ceramic composites require high temperatures but low pressures. In contrast, metallic ad organic composites require the opposite of ceramic parameters. The synthesis of organic polymers is a challenge, because they are easily decomposed at high-temperature treatments [76].

Despite hot pressing technique is widely applied to the synthesis of ceramic derivatives; it is also possible to obtain composites from other materials such as metals. Güler et al. reported developing a copper composite reinforced with aluminum oxide (Al_2O_3) and coated with silver using the hot pressing method. This synthesis was performed at 300 MPa and 250 °C. The resulting materials exhibited higher wear resistance and hardness with a higher percentage of Al_2O_3 content and improved electrical and conductive properties [115]. Kim et al. synthesized a carbon black polyamide composite through the hot pressing technique at 400 MPa and 250 °C of pressure and temperature. These materials showed better mechanical properties and suitable electromagnetic interference parameters appropriate for aerospace applications [115].

4.1.5 Atomic layer deposition technique (ALD)

The atomic layer deposition (ALD) technique creates thin films on the substrate surface (temperature, pressure, and other parameters are controlled) [116]. Two precursors must be supplied in sequence, one time, into a deposition zone [117], forming the composite film as is shown in Figure 7. The ADL precursors have pulsed alternately, one at a time, and separated by inert gas [118]. It is possible to repeat more the once to create a thicker film.

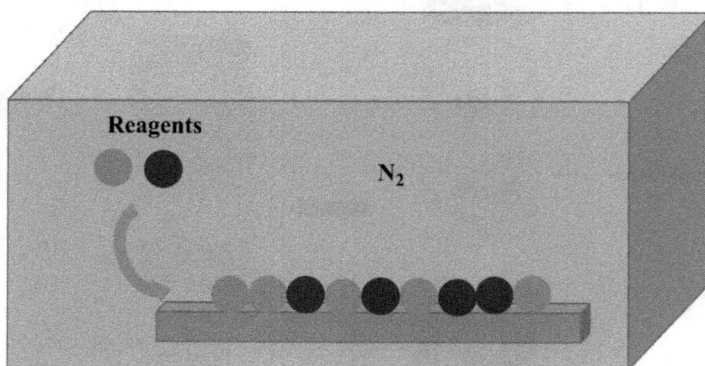

Figure 7. Vacuum chamber for atomic deposition layer.

Thermoelectric Polymers: Properties and Applications   Materials Research Forum LLC
Materials Research Foundations 162 (2024) 24-55   https://doi.org/10.21741/9781644903018-2

The applications of material synthesized through this process include inorganic and organic-based composites. One common strategy for constructing composite structures is the deposition of nanoparticles on substrates [119]. Therefore, based on that principle, Zhang et al. synthesized a polypropylene (PP) composite with titanium oxide (TiO$_2$) (hydrophilic) through the ADL technique [120].

Inorganic compounds based on metals have been studied for catalysts. Guo et al., 2020 developed a TiN@Co5.47N (cobalt nitride) material with a potential catalytic response in any reaction medium and long durability. Besides, it was reported to reuse this catalyst for up to 1500 hours [121]. Jianguo Li et al. reported the deposition of TiO$_2$ on an activated carbon surface. These compounds proved effective for removing organic contaminating agents. The deposition process was repeated three times to ensure that the entire activated carbon surface was covered [122]. Composites based on ADL synthesis are also employed in the sensor field. Zhao et al. reported a composite made of cobalt-doped porous carbon film applied as an enzyme-free lactic acid sensor for detection and high sensitivity [123]. Organic material also can be deposited onto matrices to form thin films. Wang et al. developed a composite containing pyromellitic dianhydride and ethylenediamine to produce the polyimide. Thin hybrid films were fabricated, exhibiting a permeability of 63% of water [124].

4.1.6 Three-Dimensional (3D) printing

In recent years, the fabrication of composite structures using 3D printing techniques has emerged because of the technologies' benefits [125]. 3D printing is also known as additive manufacturing, which is a technology that encompass materials, structures, and functions [86]. It works on the principle of laying down successive polymer layers to form the 3D scaffold [126]. The layer-by-layer fabrication technique enhance the creation of complex shapes and functional materials as reducing material residues [127]. Thermoplastics and their composites are used raw materials for 3D printing processes [128]. The main advantages of additive manufacturing are rapid prototyping and printability of complex entities [129]. Due to its remarkable characteristics, 3D printing is widely applied in many fields, especially in biomedical applications. Tissue engineering needs to fabricate complex multi-material biomimetic structures that resemble the in vivo tissue architecture [130]. Ni et al. reported a composite containing organic resin doped with inorganic long persistent phosphors layer by layer [131]. Shah et al. developed a hybrid organic-inorganic composite. They use a 3D printing of acrylonitrile butadiene styrene polymer doped with carbon fibers using a thermos-plastic mill for sensors, robotics, and electronics applications [132].

4.2   Characterization of organic-inorganic composites

Polymer composites are a synergic incorporation of inorganic filler into an organic matrix. The filler provides particular changes in the properties of the polymers, enhancing their characteristics. Therefore, polymer-composites are determined using mechanical, thermal and physic-chemical characterization techniques.

4.2.1 Mechanical

Mechanical characterization determines the strength and resistance of polymeric composites by material deformation. Young's modulus or also known as elastic modulus is measured to determine how much a material resists deformation [133]. In other words, the elastic modulus is the resistance exerted on the measured fabric, gradually increasing the tension and measuring the elongation until the material breaks [134]. Figure 8 shows a typical stress-strain curve that represents and compares the mechanical properties of different kinds of polymers (brittle, ductile and elastomer behavior). The stress-strain curves look different for compression and extension loadings [135]. Flexural strength is also called bend strength and modulus of rupture [136]. This method apply energy to composites, determining their resistance to bending deflection [137]. Surface modification of fibers is often required to achieve maximum compatibility fiber-matrix interfaces [138,139].

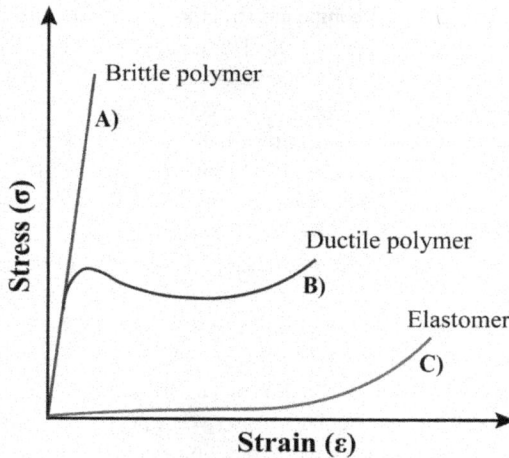

Figure 8. Stress-strain curve.

Remarkable mechanical properties characterize organic-inorganic composites. Composite materials are heterogeneous materials that often show significant advantages of stiffness and strength over homogeneous material formulations [140]. Dispersed filler particles act as a mechanical interlocking between the fiber and matrix [141] of composites. Therefore, the fillers improve either the mechanical and thermal properties of the composites [142]. Besides, when reinforcing fibers are incorporated into an organic matrix, it increases stiffness, impact strength, and tensile strength, obtaining the best mechanical test results [143].

4.2.2 Thermal

Thermal properties of organic/inorganic composites play an important role in thermal interface materials [144]. Highly cross-linking polymers tend to no melt and low cross-linking polymers increases the T_g values, restricting the chain mobility [145]. Besides, higher molecular weight of composite is more resistance to thermal decay [146,147]. Some techniques are used to study the thermal behavior of polymers such as Differential scanning calorimetry (DSC) and Thermogravimetric Analysis (TGA).

DSC is a thermal technique where the temperature of a sample and a reference is measured as a function of temperature increment [148]. DSC employs an inert atmosphere like nitrogen with a constant heat flow rate [149]. Besides, DSC determine the T_g and T_m of composites. T_g determines the reversible transition in amorphous materials from a brittle state to a viscous state [150]. Partly crystalline polymers may be flexible at temperatures above the T_g and below the T_m [151]. On other hand, TGA is one of the other routine methods for polymer [152], nanomaterials and polymer composites characterization [153]. TGA determines the weight change that occurs as a sample is heated as the temperature rise [154]. TGA reflects physical and chemical reactions including absorption, desorption, decomposition, oxidation, reduction, among others [155].

4.2.3 Microscopy

Scanning electron microscope (SEM) serves for imaging the material's microstructure and morphology [156]. SEM obtain images of composites reinforced with different types of fibers [157] due to the combination of higher magnification, greater resolution, and ease of sample observation [158]. Another interesting technique is Transmission electron microscopy (TEM). TEM characterizes the structural and chemical information of a sample at the nanometer scale [159], due to the difference in electron density of the structures in the sample [160]. TEM displays a composite material image after finishing mapping the chemical composition using an energy dispersion spectroscope [160]. Other technique also employed to study the organic-inorganic composites is the X-ray diffraction (XRD) techniques. XRD serves to distinguish between amorphous and crystalline material [161]. The crystalline phases, structural properties, and orientations by measuring the atomic arrangement of small crystal or grain regions and the thickness of thin-film layers [162, 163].

Conclusion

Polymers are interesting materials applied in many applications fields. Organic polymers are based on a primary backbone of carbon atoms. Meanwhile, inorganic polymers are based on a primary backbone distinct from carbon, like silica. These materials have been extensively studied for multiple purposes.

The polymer can be rearranged into a 3D cross-linking structure known as a polymeric matrix. This polymeric matrix is a continuous phase for incorporating other materials to improve or enhance their properties. Combining a polymeric matrix and a filler is known as a composite. Composites are made up of at least two different materials. Usually, these composites consist of an organic polymeric matrix and inorganic filler as a reinforcement. Multiple techniques have been developed to create new composites. This chapter has detailed the main techniques for fabricating organic-inorganic composites so far. Each of these techniques offers some advantages and disadvantages, based on the approach of reagents, the precursors, the process of fabrication, and some other parameters such as temperature, and pressure, among others. The electrospinning technique generates composites depending on the voltage and viscosity of the polymer precursors. Solution processing forms organic-inorganic composites based on a solvent system of polymers and an interlayer of silicate mixed. Hydrothermal, spray coating, and inkjet printing are also based on a solution principle system. Hot pressing is a temperature and pressure-dependent technique to fabricate homogeneous composite through a uniaxial application of constant pressure over two different materials. The constant pressure increases the contact points between materials. ALD is a technique carried out in a vacuum that consists of a controlled deposition of precursors into a substrate surface, pulsed alternately and separated by an inert gas, forming a composite film. The three-dimensional technique is a similar technique to ALD. Its principle relies on the successive layer deposition, forming a 3D composite structure.

Characterization techniques determine the main characteristics and properties of the final composites. The mechanical and thermal properties improved when inorganic fillers were incorporated into the organic matrix. These composites are stronger and thermal resistant compared to those cross-linking materials without additional fillers. Mechanical characterization determines the mechanical behavior of the composite. When a load is supplied, the composite is subjected to deformation, compression, or extension (flexural strength). Thermal characterization determine the thermal transition of the composites as the temperature changes. DSC determines the Tg and Tm and compares these temperatures with the composites precursors. However, TGA determines any composite's thermal stability and degradation as the temperature increases. Finally, microscopy techniques, including SEM, TEM, and XRD, allow for characterizing a composite's structure and morphology.

Studying new eco-friendly techniques to produce composite materials that fulfill the green chemistry is essential. The energy consumption must be reduced, and the precursors must be used 100%.

References

[1] B.A. Vera, L. Bastidas, M.B. Torres, P. Pinos, E. Bucio, Hyper-crosslinked Polymers, in: Inamuddin, M.I. Ahamed, R. Boddula (Eds.), Porous Polymer Science and Applications, 2022, CRC Press, pp. 7-36. https://doi.org/10.1201/9781003169604-2

[2] V. Vasiliev, E. Morozov, Advanced Mechanics of Composite Materials and Structures, Elsevier, 2018. https://doi.org/10.1016/B978-0-08-102209-2.00002-5

[3] V. Karbhari, Durability of Composites for Civil Structural Applications, CRC Press, 2007. https://doi.org/10.1533/9781845693565

[4] A. Afzal, Y. Nawab, Polymer composites, in Y. Nawab, S.M. Sapuan, K. Shaker (Eds.), Composite Solutions For Ballistics, Elsevier, 2021, pp. 139-152. https://doi.org/10.1016/B978-0-12-821984-3.00003-6

[5] A. Boccaccini, J. Blaker, Bioactive composite materials for tissue engineering scaffolds, Expert Review of Medical Devices 2 (2005) 03-317. https://doi.org/10.1586/17434440.2.3.303

[6] M. Reddy, D. Ponnamma, R. Choudhary, K. Sadasivuni, A comparative review of natural and synthetic biopolymer composite scaffolds, Polymers 13 (2021) 1105. https://doi.org/10.3390/polym13071105

[7] G. Markovic, P. Visakh, Recent Developments in Polymer Macro, Micro and Nano Blends, Woodhead Publishing, 2017.

[8] X. Sun, T. Zhang, H. Wang, Hemicelluloses-based hydrogels, in: T.K. Giri, B. Ghosh (Eds.), Plant and Algal Hydrogels for Drug Delivery and Regenerative Medicine, 2021, 181-216. https://doi.org/10.1016/B978-0-12-821649-1.00014-3

[9] X. Pei, K. Friedrich, Reference Module in Materials Science and Materials Engineering, Elsevier, 2016.

[10] M. Rahman, S. Hamdan, Study on physical, mechanical, morphological and thermal properties of styrene- co -glycidyl methacrylate/fumed silica/clay nanocomposites, in R. Rahman (Eds.), Silica and Clay Dispersed Polymer Nanocomposites, Elsevier, 2018, 71-85. https://doi.org/10.1016/B978-0-08-102129-3.00006-3

[11] J. Kenny, L. Nicolais, Comprehensive Polymer Science and Supplements, 1989, Pergamon.

[12] G. Tian, G. Han, F. Wang, J. Liang, J. Sepiolite nanomaterials: Structure, properties and functional applications, in A. Wang, W. Wang (Eds.), Nanomaterials from Clay Minerals, Elsevier, 2019, pp. 135-201. https://doi.org/10.1016/B978-0-12-814533-3.00003-X

[13] M. Tanzi, S. Farè, G. Candiani, Foundations of Biomaterials Engineering, Elsevier, 2019.

[14] G. Scheutz, J. Lessard, M. Sims, B. Sumerlin, Adaptable crosslinks in polymeric materials: Resolving the intersection of thermoplastics and thermosets, Journal of The American Chemical Society 141 (2019) 16181-16196. https://doi.org/10.1021/jacs.9b07922

[15] V. Kulkarni, C. Shaw, Essential Chemistry for Formulators of Semisolid and Liquid Dosages, Academic Press, 2016.

[16] M. Soroush, M. Grady, Polymers, Polymerization Reactions, and Computational Quantum Chemistry, in: J. Izaac, J. Wang (Eds.), Computational Quantum Chemistry, Springer International Publishing, 2019, pp. 1-16. https://doi.org/10.1016/B978-0-12-815983-5.00001-5

[17] C. Sarathchandran, Rheology of Polymer Blends and Nanocomposites, Elsevier, 2020. https://doi.org/10.1016/B978-0-12-816957-5.00006-9

[18] M.B. Torres, V.P. Ramos, D.R. Fierro, S.H. Bonilla, H. Magaña, E. Bucio, Synthesis and antimicrobial properties of highly cross-linked pH-sensitive hydrogels through gamma radiation, Polymers 13 (2021) 2223. https://doi.org/10.3390/polym13142223

[19] B. Bolto, Z. Xie, The use of polymers in the flotation treatment of wastewater, Processes 7 (2019) 374. https://doi.org/10.3390/pr7060374

[20] S. Guilbert, P. Feuilloley, H. Bewa, V. Bellonmaurel, Biodegradable polymers in agricultural applications, in: R. Smith (Eds.), Biodegradable Polymers for Industrial Applications, Woodhead Publishing, 2005, pp. 494-516. doi: 10.1533/9781845690762.4.494 https://doi.org/10.1533/9781845690762.4.494

[21] H. Brinson, L. Brinson, Polymer Engineering Science and Viscoelasticity, Springer New York, 2015. https://doi.org/10.1007/978-1-4899-7485-3

[22] N. Cheremisinoff, Condensed Encyclopedia of Polymer Engineering Terms, Butterworth-Heinemann, 2001, pp. 165-182. https://doi.org/10.1016/B978-0-08-050282-3.50018-4

[23] J. Aklonis, Mechanical properties of polymers, Journal of Chemical Education 58 (1981) 892. https://doi.org/10.1021/ed058p892

[24] W. Su, Polymer Size and Polymer Solutions, Lecture Notes in Chemistry, 2013, pp. 9-26. https://doi.org/10.1007/978-3-642-38730-2_2

[25] R. Hill, Polymers. Biomaterials, Artificial Organs and Tissue Engineering, Woodhead Publishing, 2005. https://doi.org/10.1201/9780203024065.ch4

[26] D. Jones, M. Ashby, Mechanisms of Creep, and Creep-Resistant Materials. Engineering Materials 1 (2019) 381-394. https://doi.org/10.1016/B978-0-08-102051-7.00022-1

[27] F. Qiu, Accelerated Predictive Stability, Academic Press, 2018. https://doi.org/10.1016/B978-0-12-802786-8.00001-2

[28] D. Eskin, Ultrasonic degassing of liquids, in: J.A.G. Juárez, K.F. Graff (Eds.), Power Ultrasonics, Woodhead Publishing, 2015, 611-631. https://doi.org/10.1016/B978-1-78242-028-6.00020-X

[29] M. Khan, A. Svedberg, A. Singh, M. Ansari, Z. Karim, Nanostructured Polymer Composites for Biomedical Applications, Elsevier, 2019.

[30] S. Gad, Encyclopedia of Toxicology, Academic Press, 2014. https://doi.org/10.1016/B978-0-12-386454-3.00823-X

[31] A. Yadav, N. Sinha, Organic Polymers for Drinking Water Purification. Reference Module In Materials Science And Materials Engineering, 2021. https://doi.org/10.1016/B978-0-12-820352-1.00140-1

[32] C. Donga, K. Mabape, S. Mishra, A. Mishra, Polymer-based engineering materials for removal of nano wastes from water, in: A.K. Mishra, H.M.D. Anawar, N. Drouiche (Eds.), Emerging and Nanomaterial Contaminants In Wastewater, Elsevier, 2019, pp. 217-243. https://doi.org/10.1016/B978-0-12-814673-6.00008-5

[33] A. Maiti, A. Mule, A. Kumar, A. Bhatnagar, P. Mondal, Polymers in Wastewater Treatment, Reference Module in Materials Science and Materials Engineering, 2021. https://doi.org/10.1016/B978-0-12-820352-1.00148-6

[34] P. Joseph, J. Ebdon, Recent developments in flame-retarding thermoplastics and thermosets, in: A.R. Horrocks, D. Price (Eds.), Fire Retardant Materials, Woodhead Publishing, 2001, 220-263. https://doi.org/10.1533/9781855737464.220

[35] R. Salunke, The sources of heavy metals, its impact on human life and the progress in electrochemical sensor, in: C.M. Hussain, S. Shukla, G. Joshi (Eds.), Functionalized Nanomaterials Based Devices for Environmental Applications, Elsevier, 2021, 349-378. https://doi.org/10.1016/B978-0-12-822245-4.00016-7

[36] E. Ivanova, K. Bazaka, R. Crawford, New Functional Biomaterials for Medicine and Healthcare, 2014, 100-120. https://doi.org/10.1533/9781782422662.100

[37] K. MacKenzie, Innovative applications of inorganic polymers (geopolymers), in: F. Torgal, J.A. Labrincha, C. Leonelli, A. Palomo, P. Chindaprasirt (Eds.), Handbook of Alkali-Activated Cements, Mortars, and Concretes, Woodhead Publishing, 2015, 777-805. https://doi.org/10.1533/9781782422884.5.777

[38] Z. Mohamad, S. Man, N. Othman, N. Abdullah, M. Abdulwasiu, Reference Module In Materials Science And Materials Engineering, 2021.

[39] L. Ansaloni, Advances in polymer-inorganic hybrids as membrane materials, Recent Developments in Polymer Macro, Micro and Nano Blends, 2017, 163-206. https://doi.org/10.1016/B978-0-08-100408-1.00007-8

[40] J. Brauns, Reinforced materials: Elastic Properties and Strength Prediction, Encyclopedia of Materials: Science and Technology, 1-9. https://doi.org/10.1016/B0-08-043152-6/02137-9

[41] M. Sultan, Thermal properties of oil palm biomass-based composites, in: M. Jawaid, P.M. Tahir, N. Saba (Eds.), Lignocellulosic Fibre and Biomass-Based Composite Materials, Woodhead Publishing, 95-122.

[42] D. Sharma, Materials for Biomedical Engineering, Elsevier, 2019.

[43] P. Mallick, Materials, Design and Manufacturing for Lightweight Vehicles, Woodhead Publishing, 2010. https://doi.org/10.1533/9781845697822

[44] M. Asim, M. Jawaid, N. Saba, N. Ramengmawii, M. Nasir, M. Sultan, Processing of hybrid polymer composites-a review, in: V.K. Thakur, M.K. Thakur, A. Pappu (Eds.), Hybrid Polymer Composite Materials, Woodhead Publishing, 2017, pp.1-22. https://doi.org/10.1016/B978-0-08-100789-1.00001-0

[45] H. Yahyaei, M. Mohseni, Polyhedral Oligomeric Silsesquioxane (POSS) Polymer Nanocomposites, 2021, 115-125. https://doi.org/10.1016/B978-0-12-821347-6.00013-5

[46] Yan, Y. Shi, Z. Li, Selective Laser Sintering Additive Manufacturing Technology, 2021, 667-712. https://doi.org/10.1016/B978-0-08-102993-0.00005-9

[47] T. Lee, F. Boey, K. Khor, On the determination of polymer crystallinity for a thermoplastic PPS composite by thermal analysis, Composites Science and Technology 53 (1995) 259-274. https://doi.org/10.1016/0266-3538(94)00070-0

[48] N. Yılmaz, A. Khan, Flexural behavior of textile-reinforced polymer composites, in: M. Jawaid, M. Thariq, N. Saba (Eds.), Mechanical and Physical Testing of Biocomposites, Fibre-Reinforced Composites and Hybrid Composites, Woodhead Publishing, 2019, pp. 13-42. https://doi.org/10.1016/B978-0-08-102292-4.00002-3

[49] M. Biron, Material Selection for Thermoplastic Parts, Elsevier Science, 2016. https://doi.org/10.1016/B978-0-7020-6284-1.00003-9

[50] A. Marques, Fibrous and Composite Materials for Civil Engineering Applications, Woodhead Publishing, 2011.

[51] S. Khatiwada, U. Gohs, R. Lach, G. Heinrich, R. Adhikari, A new way of toughening of thermoset by dual-cured thermoplastic/thermosetting blend. Materials 12 (2019), 548. https://doi.org/10.3390/ma12030548

[52] P. Khui, M. Rahman, E. Jayamani, Advances in Sustainable Polymer Composites, Elsevier, 2021. https://doi.org/10.1016/B978-0-12-820338-5.00012-6

[53] J. Greene, Automotive Plastics and Composites, William Andrew, 2021. https://doi.org/10.1016/B978-0-12-818008-2.00017-9

[54] S. Amir, M. Sultan, M. Jawaid, A. Ariffin, S. Mohd, K. Salleh, Nondestructive testing method for Kevlar and natural fiber and their hybrid composites, in: M. Jawaid, M. Thariq, N. Saba (Eds.), Durability and Life Prediction in Biocomposites, Fibre-Reinforced Composites and Hybrid Composites, Woodhead Publishing, 2019, 367-388. https://doi.org/10.1016/B978-0-08-102290-0.00016-7

[55] S. Ogin, P. Brøndsted, J. Zangenberg, Modeling Damage, Fatigue and Failure of Composite Materials, Elsevier, 2016. https://doi.org/10.1016/B978-1-78242-286-0.00001-7

[56] M. Knight, D. Curliss, Composite Materials, in: R.A. Meyers (Eds.), Encyclopedia of Physical Science and Technology, Academic Press, 2003, pp. 455-468. https://doi.org/10.1016/B0-12-227410-5/00128-9

[57] D. Aleksendrić, P. Carlone, Soft Computing in The Design and Manufacturing of Composite Materials, 2015, 1-5. https://doi.org/10.1533/9781782421801.1

[58] X. Sun, T. Zhang, H. Wang, Plant and Algal Hydrogels for Drug Delivery and Regenerative Medicine, Woodhead Publishing, 2021.

[59] D. Baird, Polymer processing, in: R.A. Meyers (Eds.), Encyclopedia of Physical Science and Technology, Academic Press, 2003, 611-643. https://doi.org/10.1016/B0-12-227410-5/00593-7

[60] P. Gradin, R. Seldén, R. Brown, Dynamic-mechanical Properties, in: G. Allen, S.L. Aggarwal, S. Russo (Eds.), Comprehensive Polymer Science and Supplements, Pergamon, 1989, 533-569. https://doi.org/10.1016/B978-0-08-096701-1.00053-7

[61] L. McKeen, The Effect of Long-Term Thermal Exposure On Plastics And Elastomers, Kidlington: William Andrew, 2021.

[62] J. Pojman, Polymer Science: A Comprehensive Reference, Elsevier Science, 2012, 957-980. https://doi.org/10.1016/B978-0-444-53349-4.00124-2

[63] T. Naylor, Permeation properties, in: G. Allen, S.L. Aggarwal, S. Russo (Eds.), Comprehensive Polymer Science and Supplements, Pergamon, 1989, pp. 643-668. https://doi.org/10.1016/B978-0-08-096701-1.00057-4

[64] V. Vasiliev, E. Morozov, Introduction. Advanced Mechanics of Composite Materials, Elsevier Science, 2013. https://doi.org/10.1016/B978-0-08-098231-1.00001-7

[65] A.M. Taib, N. Julkapli, Mechanical And Physical Testing Of Biocomposites, Fibre-Reinforced Composites And Hybrid Composites, 61-79.

[66] Asthana, N., & Pal, K. (2020). Polymerized hybrid nanocomposite implementations of energy conversion cells device. Nanofabrication For Smart Nanosensor Applications, 349-397. https://doi.org/10.1016/B978-0-12-820702-4.00015-5

[67] O. Adekomaya, T. Majozi, Fiber Reinforced Composites, Elsevier, 2021.

[68] Cantor, K., & Watts, P. (2011). Plastics Processing. Applied Plastics Engineering Handbook, 195-203. https://doi.org/10.1016/B978-1-4377-3514-7.10012-1

[69] H. Dhakal, S. Ismail, Introduction to composite materials, in: H.N. Dhakal S.O. Ismail (Eds.), Sustainable Composites For Lightweight Applications, Woodhead Publishing, 2021, pp. 1-16. https://doi.org/10.1016/B978-0-12-818316-8.00001-3

[70] R. Guedes, J. Xavier, Advanced Fibre-Reinforced Polymer (FRP) Composites for Structural Applications, Woodhead Publishing, 2013. https://doi.org/10.1533/9780857098641.3.298

[71] J. Alemán, A. Chadwick, J. He, M. Hess, K. Horie, R. Jones, Definitions of terms relating to the structure and processing of sols, gels, networks, and inorganic-organic hybrid materials (IUPAC Recommendations 2007), Pure and Applied Chemistry 79 (2007), 1801-1829. https://doi.org/10.1351/pac200779101801

[72] Ou, A. Sangle, A. Datta, Q. Jing, T. Busolo, T. Chalklen, Fully printed organic-inorganic nanocomposites for flexible thermoelectric applications, ACS Applied Materials &Amp. Interfaces 10 (2018) 19580-19587. https://doi.org/10.1021/acsami.8b01456

[73] B. McGrail, A. Sehirlioglu, E. Pentzer, Polymer composites for thermoelectric applications, Angewandte Chemie International Edition 54 (2014) 1710-1723. https://doi.org/10.1002/anie.201408431

[74] M. Dresselhaus, G. Chen, M. Tang, R. Yang, H. Lee, D. Wang, New directions for low-dimensional thermoelectric materials, Advanced Materials 19 (2007) 1043-1053. https://doi.org/10.1002/adma.200600527

[75] P. Palmero, Encyclopedia of Materials: Technical Ceramics And Glasses, 2021, 501-510. https://doi.org/10.1016/B978-0-12-818542-1.00013-8

[76] T. Miyazaki, K. Ishikawa, Y. Shirosaki, C. Ohtsuki, Organic-inorganic composites designed for biomedical applications, Biological and Pharmaceutical Bulletin 36 (2013) 1670-1675. https://doi.org/10.1248/bpb.b13-00424

[77] F. Mammeri, E. Bourhis, L. Rozes, C. Sanchez, Mechanical properties of hybrid organic-inorganic materials, Journal of Materials Chemistry 15 (2005) 3787. https://doi.org/10.1039/b507309j

[78] M. Virji, A. Stefaniak, A review of engineered nanomaterial manufacturing processes and associated exposures, in: S. Hashmi, G.F. Batalha, B. Yilbas (Eds.), Comprehensive Materials Processing, Elsevier, 2014, pp.103-125. https://doi.org/10.1016/B978-0-08-096532-1.00811-6

[79] A. Reddy, G. Reddy, V. Sivanjineyulu, J. Jayaramudu, K. Varaprasad, E. Sadiku, Design and Applications Of Nanostructured Polymer Blends And Nanocomposite Systems, William Andrew, 2016, 385-411. https://doi.org/10.1016/B978-0-323-39408-6.00016-9

[80] Wang, X. X., Yu, G. F., Zhang, J., Yu, M., Ramakrishna, S., & Long, Y. Z. (2021). Conductive Polymer Ultrafine Fibers via Electrospinning: Preparation, Physical Properties and Applications. Progress in Materials Science, 115, 100704. https://doi.org/10.1016/j.pmatsci.2020.100704

[81] D. Ficai, M. Albu, Advances in the field of soft tissue engineering. Nanobiomaterials In Soft Tissue Engineering, William Andrew, 2016, 355-386. https://doi.org/10.1016/B978-0-323-42865-1.00013-1

[82] Y. Li, J. Zhu, H. Cheng, G. Li, H. Cho, M. Jiang, Q. Gao, X. Zhang, Developments of advanced electrospinning techniques: A critical review. Advanced Materials Technologies, 6 (2021) 2100410. https://doi.org/10.1002/admt.202100410

[83] E. Uslu, Determination of mechanical properties of polymer matrix composites reinforced with electrospinning N66, PAN, PVA, and PVC nanofibers: A comparative study, Materials Today Communications 26 (2021) 101939. https://doi.org/10.1016/j.mtcomm.2020.101939

[84] J. Cui, Flexible and transparent composite nanofibre membrane that was fabricated via a "green" electrospinning method for efficient particulate matter 2.5 capture, Journal of Colloid and Interface Science 582 (2021) 506-514. https://doi.org/10.1016/j.jcis.2020.08.075

[85] K. Peranidze, T.V. Safronova, Fibrous polymer-based composites obtained by electrospinning for bone tissue engineering, Polymers 14 (2022) 1-19. https://doi.org/10.3390/polym14010096

[86] J. Yang, N. Li, J. Shi, W. Tang, G. Zhang, Introduction. Multimaterial 3D Printing Technology, 2021. https://doi.org/10.1016/B978-0-08-102991-6.00014-8

[87] R. Zhang, Development of polylactic acid/ZnO composite membranes prepared by ultrasonication and electrospinning for food packaging, Lwt 135 (2021) 110072. https://doi.org/10.1016/j.lwt.2020.110072

[88] C.L. McCarthy, R.L. Brutchey, Solution processing of chalcogenide materials using thiol-amine "alkahest" solvent systems, Chemical Communications 53 (2017), 4888-4902. https://doi.org/10.1039/C7CC02226C

[89] A.V. Rane, K. Kanny, V. Abitha, S. Thomas, Synthesis of Inorganic Nanomaterials, Woodhead Publishing, 2018.

[90] A.V. Rane, K. Kanny, V. Abitha, S. Patil, S. Thomas, Clay-Polymer Composites, in: K. Jlassi, M.M. Chehimi, S. Thomas (Eds.), Clay-Polymer Nanocomposites, Elsevier, 2017, 113-144. https://doi.org/10.1016/B978-0-323-46153-5.00004-5

[91] L. Tang, L. Zhao, F. Qiang, Q. Wu, L. Gong, J. Peng, Mechanical Properties of Rubber Nanocomposites Containing Carbon Nanofillers, in: S. Yaragalla (Eds.), Carbon-Based Nanofillers and Their Rubber Nanocomposites, Elsevier, 2019, 367-423. https://doi.org/10.1016/B978-0-12-817342-8.00012-3

[92] Deshmukh, K., Basheer Ahamed, M., Deshmukh, R., Khadheer Pasha, S., Bhagat, P., & Chidambaram, K. (2017). Biopolymer Composites with High Dielectric Performance: Interface Engineering. Biopolymer Composites in Electronics, 27-128. https://doi.org/10.1016/B978-0-12-809261-3.00003-6

[93] S. Feng, G. Li, Modern Inorganic Synthetic Chemistry, 2017.

[94] Z. Xu, C. Gao, In situ polymerization approach to graphene-reinforced nylon-6 composites, Macromolecules 43 (2010), 6716-6723. https://doi.org/10.1021/ma1009337

[95] Y. Chao, Y. Ge, Z. Chen, X. Cui, C. Zhao, C. Wang, G.G. Wallace, One-pot hydrothermal synthesis of solution-processable MoS2/PEDOT: PSS composites for high-performance supercapacitors, ACS Applied Materials and Interfaces, 13 (2021) 7285-7296. https://doi.org/10.1021/acsami.0c21439

[96] F.S. Kordshuli, F. Zabihi, Graphene-doped PEDOT: PSS nanocomposite thin films fabricated by conventional and substrate vibration-assisted spray coating (SVASC), Engineering Science and Technology 19 (2016) 1216-1223. https://doi.org/10.1016/j.jestch.2016.02.003

[97] A. Elzoghby, M. Elgohary, N. Kamel, Implications of protein- and peptide-based nanoparticles as potential vehicles for anticancer drugs, in: R. Donev (Eds.), Advances in Protein Chemistry and Structural Biology, Academic Press, 2015, 169-221. https://doi.org/10.1016/bs.apcsb.2014.12.002

[98] Ramos Avilez, H., Castilla Casadiego, D., Vega Avila, A., Perales Perez, O., & Almodovar, J. (2017). Production of chitosan coatings on metal and ceramic biomaterials. Chitosan Based Biomaterials Volume 1, 255-293. https://doi.org/10.1016/B978-0-08-100230-8.00011-X

[99] J. McCollum, S. Delgado, Manufacturing strategies in fluorinated polymers and composites, in: B. Ameduri, S. Fomin (Eds,), Opportunities for Fluoropolymers, Elsevier, 2020, 275-301. https://doi.org/10.1016/B978-0-12-821966-9.00010-9

[100] C. Cie, Theoretical foundations for inkjet technology, Ink Jet Textile Printing, 2015, 1-13. https://doi.org/10.1016/B978-0-85709-230-4.00001-7

[101] J. Perelaer, U. Schubert, Ink-Jet Printing of Functional Polymers for Advanced Applications, in: K. Matyjaszewski, M. Möller (Eds.), Polymer Science: A Comprehensive Reference, Elsevier, 2012, pp. 147-175. https://doi.org/10.1016/B978-0-444-53349-4.00205-3

[102] E. Loser, H. Tobler, Digital Printing of Textiles, Woodhead Publishing, 2006.

[103] H. Kobayashi, Industrial production printers - Mimaki's Tx series, in: H Ujiie (Eds.), Digital Printing of Textiles, Woodhead Publishing, 2006, pp. 98-122. https://doi.org/10.1533/9781845691585.1.98

[104] A.S. Gorgani, Inkjet Printing, in: J. Izdebska, S. Thomas (Eds), Printing on Polymers, William Andrew, 2016, 231-246. https://doi.org/10.1016/B978-0-323-37468-2.00014-2

[105] J. Zhang, K. Hoshino, Fundamentals of Nano/Microfabrication and Effect of Scaling, in: J.X.J. Zhang, K. Hoshino (Eds.), Molecular Sensors and Nanodevices, Academic Press, 2014, pp. 43-101. https://doi.org/10.1016/B978-1-4557-7631-3.00002-8

[106] S. Ko, Advanced Inkjet Technology for 3D Micro-metal Structure Fabrication, in: Y. Qin (Eds.), Micromanufacturing Engineering and Technology, William Andrew, 2015, 425-439. https://doi.org/10.1016/B978-0-323-31149-6.00018-9

[107] K. Kato, H. Hagino, K. Miyazaki, Fabrication of bismuth telluride thermoelectric films containing conductive polymers using a printing method, Journal of Electronic Materials 42 (2013) 1313-1318. https://doi.org/10.1007/s11664-012-2420-z

[108] M. Mikolajek, T. Reinheimer, N. Bohn, C. Kohler, M. Hoffmann, J. Binder, Fabrication and characterization of fully inkjet printer capacitors based on ceramic/polymer composite dielectrics on flexible substrates. Scientific Reports 9 (2019) 1334. https://doi.org/10.1038/s41598-019-49639-3

[109] M. Klein, S. Steenhusen, P. Löbmann, Inorganic-organic hybrid polymers for printing of optical components: From digital light processing to inkjet 3D-printing, Journal of Sol-Gel Science and Technology 101 (2021) 649-654. https://doi.org/10.1007/s10971-022-05748-6

[110] J. Hostaša, Ceramics for Laser Technologies. Encyclopedia Of Materials: Technical Ceramics and Glasses, 2021, 110-124. https://doi.org/10.1016/B978-0-12-803581-8.11779-5

[111] J. Binner, Encyclopedia of Materials: Technical Ceramics and Glasses, 2021, 3-24. https://doi.org/10.1016/B978-0-12-818542-1.00067-9

[112] R. German, Sintering with External Pressure. Sintering: From Empirical Observations to Scientific Principles, 2014, 305-354. https://doi.org/10.1016/B978-0-12-401682-8.00010-0

[113] K. Shanmugam, Nanocellulose and its composite films: Applications, properties, fabrication methods, and their limitations. Nanoscale Processing, 2021, 247-297. https://doi.org/10.1016/B978-0-12-820569-3.00010-4

[114] B. Song, Microstructural characterization of TiB2-SiC-BN ceramics prepared by hot pressing. Ceramics International, 47 (2021) 29174-29182. https://doi.org/10.1016/j.ceramint.2021.07.080

[115] O. Güler, T. Varol, U. Alver, G. Kaya, F. Yıldız, Microstructure and wear characterization of Al2O3 reinforced silver coated copper matrix composites by electroless plating and hot-pressing methods, Materials Today Communications 27 (2021) 102205. https://doi.org/10.1016/j.mtcomm.2021.102205

[116] J. Hendrickson, A. Homyk, A. Scherer, T. Alasaarela, A. Säynätjoki, S. Honkanen, Quantum Optics with Semiconductor Nanostructures, Elsevier Science, 2012.

[117] S. Cho, M. Uddin, P. Alaboina, Review of nanotechnology for cathode materials in batteries, in: L.M.R. Martinez, N. Omar (Eds.), Emerging Nanotechnologies in Rechargeable Energy Storage Systems, Elsevier, 2017, pp. 83-129. https://doi.org/10.1016/B978-0-323-42977-1.00003-0

[118] M. Leskelä, J. Niinistö, M. Ritala, Comprehensive Materials Processing, Elsevier, 2014.

[119] P. Oviroh, R. Akbarzadeh, D. Pan, R. Coetzee, T. Jen, New development of atomic layer deposition: processes, methods and applications, Science And Technology Of Advanced Materials 20 (2019), 465-496. https://doi.org/10.1080/14686996.2019.1599694

[120] L. Zhang, Y. Feng, Y. Li, Y. Jiang, S. Wang, J. Xiang, J. Zhang, P. Cheng, N. Tang, Stable construction of superhydrophobic surface on polypropylene membrane via atomic layer deposition for high salt solution desalination, Journal of Membrane Science 647 (2022) 120289. https://doi.org/10.1016/j.memsci.2022.120289

[121] D. Guo, Z. Wan, Y. Li, B. Xi, C. Wang, TiN @ Co5.47N composite material constructed by atomic layer deposition as reliable electrocatalyst for oxygen evolution reaction, Advanced Functional Materials 31 (2020) 2008511. https://doi.org/10.1002/adfm.202008511

[122] J. Li, L. Hui, W. Zhang, J. Lu, Y. Yang, H. Feng, Scalable production of ultra-small TiO2 nanocrystal / activated carbon composites by atomic layer deposition for efficient removal of organic pollutants, Advanced Powder Technology 32 (2021) 728-739. https://doi.org/10.1016/j.apt.2021.01.013

[123] Z. Zhao, Y. Kong, C. Liu, G. Huang, Z. Xiao, H. Zhu, Z. Bao, Y. Mei, Atomic layer deposition-assisted fabrication of 3D Co-doped carbon framework for sensitive enzyme-free lactic acid sensor, Chemical Engineering Journal 417 (2021) 129285. https://doi.org/10.1016/j.cej.2021.129285

[124] H. Wang, M. Wei, Z. Zhong, Y. Wang, Atomic-layer-deposition-enabled thin-film composite membranes of polyimide supported on nanoporous anodized alumina,

Journal of Membrane Science 535 (2017) 56-62.
https://doi.org/10.1016/j.memsci.2017.04.026

[125] M. Somireddy, Fabrication of composite structures via 3D printing, in: J.P. Davim (Eds.), Materials Forming, Machining and Tribology, Springer, 2021, pp. 255-276. https://doi.org/10.1007/978-3-030-68024-4_14

[126] M. Kun, C. Chan, S. Ramakrishna, A. Kulkarni, K. Vadodaria, Textile-based scaffolds for tissue engineering, in: S. Rajendran (Eds.), Advanced Textiles for Wound Care, Woodhead Publishing, 2019, pp. 329-362. https://doi.org/10.1016/B978-0-08-102192-7.00012-6

[127] G. Goh, S. Sing, W. Yeong, A review on machine learning in 3D printing: Applications, potential, and challenges, Artificial Intelligence Review 54 (2020) 63-94. https://doi.org/10.1007/s10462-020-09876-9

[128] X. Tian, A. Todoroki, T. Liu, L. Wu, Z. Hou, M. Ueda, 3D printing of continuous fiber reinforced polymer composites: Development, application, and perspective, Chinese Journal of Mechanical Engineering: Additive Manufacturing Frontiers 1 (2022) 100016. https://doi.org/10.1016/j.cjmeam.2022.100016

[129] Q. Jiang, J. Yang, P. Hing, H. Ye, Recent advances, design guidelines, and prospects of flexible organic/inorganic thermoelectric composites, Materials Advances, 1 (2020) 1038-1054. https://doi.org/10.1039/D0MA00278J

[130] C. Vyas, G. Poologasundarampillai, J. Hoyland, P. Bartolo, 3D printing of biocomposites for osteochondral tissue engineering, in: L. Ambrosio (Eds.), Biomedical Composites, Woodhead Publishing, 2017, pp. 261-302. https://doi.org/10.1016/B978-0-08-100752-5.00013-5

[131] R. Ni, B. Qian, C. Liu, X. Liu, J. Qiu, Three-dimensional printing of hybrid organic/inorganic composites with long persistence luminescence, Optical Materials Express, 8 (2018) 2823. https://doi.org/10.1364/OME.8.002823

[132] S. Shah, M. Shiblee, J. Rahman, S. Basher, S. Mir, M. Kawakami, 3D printing of electrically conductive hybrid organic-inorganic composite materials, Microsystem Technologies 24 (2018) 4341-4345. https://doi.org/10.1007/s00542-018-3781-x

[133] Kadkhodaie, A., & Kadkhodaie, R. (2022). Acoustic, density, and seismic attribute analysis to aid gas detection and delineation of reservoir properties. Sustainable Geoscience For Natural Gas Subsurface Systems, 51-92. https://doi.org/10.1016/B978-0-323-85465-8.00007-8

[134] R. Landel, L. Nielsen, Mechanical Properties of Polymers and Composites, CRC Press, 1993 https://doi.org/10.1201/b16929

[135] N. Morita, Finite Element Programming In Nonlinear Geomechanics And Transient Flow, Elsevier Science & Technology, 2021. https://doi.org/10.1016/B978-0-323-91112-2.00004-5

[136] B.V. Ramnath, V. Manickavasagam, S. Rajesh, A. Khan, A. Asiri, H.D. Cancar, Behavior of some natural fiber composites, in : A. Khan, S.M. Rangappa, S. Siengchin,

M. Jawaid, A.M. Asiri (Eds.), Hybrid Natural Fiber Composites, Woodhead Publishing, 2021, pp, 167-183.

[137] T. Pan, S. Mondal, Structural Properties and Sensing Characteristics of Sensing Materials. Comprehensive Materials Processing, (2014) 179-203. https://doi.org/10.1016/B978-0-08-096532-1.01306-6

[138] B. Yousif, A. Shalwan, C. Chin, K. Ming, Flexural properties of treated and untreated kenaf/epoxy composites, Materials &Amp; Design, 40 (2012) 378-385. https://doi.org/10.1016/j.matdes.2012.04.017

[139] A. Benkhelladi, H. Laouici, A. Bouchoucha, Tensile and flexural properties of polymer composites reinforced by flax, jute and sisal fibres, The International Journal Of Advanced Manufacturing Technology 108 (2020) 895-916. https://doi.org/10.1007/s00170-020-05427-2

[140] R. Christensen, (1983). Mechanical properties of composite materials. in: A. Kaw (Eds.), Mechanics of Composite Materials, 1997, pp. 1-16. https://doi.org/10.1016/B978-0-08-029384-4.50008-0

[141] M. Kamal, Scanning Electron Microscopy Study of Fiber Reinforced Polymeric Nanocomposites. Scanning Electron Microscopy, Intech Open, 2012. https://doi.org/10.5772/35494

[142] Z. Xu, C. Gao, In situ polymerization approach to graphene-reinforced nylon-6 composites, Macromolecules 43 (2010) 6716-6723. https://doi.org/10.1021/ma1009337

[143] Z. Ghalib, A.Al Jlaihawi, S. Ghani, A. Taher, Mechanical properties of composite materials epoxy/fiberglass/rubber, Journal Of Mechanical Engineering Research And Developments 24 (2021) 215-221

[144] B. Liu, L. Dong, Q. Xi, X. Xu, J. Zhou, B. Li, Thermal transport in organic/inorganic composites, Frontiers In Energy 12 (2018) 72-86. https://doi.org/10.1007/s11708-018-0526-6

[145] R. Seymour, C. Carraher, Thermal properties of polymers, in: R. Seymour, C. Carraher (Eds.), Structure-Property Relationships in Polymers, Springer, New York, 1984, pp. 83-93. https://doi.org/10.1007/978-1-4684-4748-4_7

[146] M. Umar, M. Ofem, A. Anwar, A. Salisu, Thermo gravimetric analysis (TGA) of PA6/G and PA6/GNP composites using two processing streams, Journal of King Saud University - Engineering Sciences, 34 (2022) 77-87. https://doi.org/10.1016/j.jksues.2020.09.003

[147] Z. Ali, Y. Gao, B. Tang, X. Wu, Y. Wang, M. Li, Preparation, properties and mechanisms of carbon fiber/polymer composites for thermal management applications, Polymers 13 (2021), 169. https://doi.org/10.3390/polym13010169

[148] R. Karoui, Chemical Analysis of Food: Techniques And Applications, Elsevier, 2012.

[149] M.B. Torres, V.P. Ramos, D.R. Fierro, S.H. Bonilla, H.S. Magaña, E. Bucio, Synthesis and antimicrobial properties of highly cross-linked ph-sensitive hydrogels through gamma radiation, Polymers 13 (2021) 2223. https://doi.org/10.3390/polym13142223

[150] B. Tomoda, P.Y. Cordeiro, J. Ernesto, P. Lopes, L. Péres, L., C. da Silva, M. de Moraes, Characterization of biopolymer membranes and films: Physicochemical, mechanical, barrier, and biological properties, Biopolymer Membranes and Films (2020) 67-95. https://doi.org/10.1016/B978-0-12-818134-8.00003-1

[151] D. Bolcu, M. Stănescu, A study of the mechanical properties of composite materials with a dammar-based hybrid matrix and two types of flax fabric reinforcement, Polymers 12 (2020) 1649. https://doi.org/10.3390/polym12081649

[152] T. Inan, Recent Developments in Polymer Macro, Micro and Nano Blends, Elsevier Science, 2017.

[153] S. Loganathan, R. Valapa, R. Mishra, G. Pugazhenthi, S. Thomas, Thermal and Rheological Measurement Techniques for Nanomaterials Characterization, Elsevier, 2017.

[154] K. Rajisha, B. Deepa, L. Pothan, S. Thomas, Interface Engineering of Natural Fibre Composites For Maximum Performance, Woodhead Publishing in Materials, pp. 241-274.

[155] J. Yang, N. Hedin, Low Carbon Stabilization and Solidification of Hazardous Wastes, Elsevier, 2022. https://doi.org/10.1016/B978-0-12-824004-5.00020-7

[156] M. Omidi, A. Fatehinya, M. Farahani, Z. Akbari, S. Shahmoradi, & F. Yazdian, Characterization of biomaterials, Biomaterials for Oral and Dental Tissue Engineering (2017) 97-115. https://doi.org/10.1016/B978-0-08-100961-1.00007-4

[157] B. Sabuncuoglu, S. Orlova, L. Gorbatikh, S. Lomov, & I. Verpoest, Micro-scale finite element analysis of stress concentrations in steel fiber composites under transverse loading, Journal of Composite Materials 49 (2014) 1057-1069. https://doi.org/10.1177/0021998314528826

[158] N. Altawell, Introduction to Machine Olfaction Devices, Academic Press, 2022. https://doi.org/10.1016/B978-0-12-822420-5.00004-0

[159] Q. Wu, M. Li, Y. Gu, S. Wang, & Z. Zhang, Imaging the interphase of carbon fiber composites using transmission electron microscopy: Preparations by focused ion beam, ion beam etching, and ultramicrotomy, Chinese Journal of Aeronautics 28 (2015) 1529-1538. https://doi.org/10.1016/j.cja.2015.05.005

[160] L. Dobrzański, B. Tomiczek, M. Pawlyta, & P. Nuckowski, TEM and XRD Study of nanostructured composite materials reinforced with the halloysite particles, Materials Science Forum 783-786 (2014) 1591-1596. https://doi.org/10.4028/www.scientific.net/MSF.783-786.1591

[161] M. Nasrollahzadeh, M. Atarod, M. Sajjadi, S. Sajadi, & Z. Issaabadi, Plant-mediated green synthesis of nanostructures: Mechanisms, characterization, and applications, Interface Science and Technology (2019) 199-322. https://doi.org/10.1016/B978-0-12-813586-0.00006-7

[162] K.E. Sapsford, Analyzing nanomaterial bioconjugates: A review of current and emerging purification and characterization techniques, Analytical Chemistry 83 (2011) 4453-4488. https://doi.org/10.1021/ac200853a

[163] B.M. Torres, D.R. -Fierro, B.A. Vera, S. Pardo, & E. Bucio, Interaction between filler and polymeric matrix in nanocomposites: Magnetic approach and applications, Polymers 13 (2021) 2998. https://doi.org/10.3390/polym13172998

Thermoelectric Polymers: Properties and Applications   Materials Research Forum LLC
Materials Research Foundations 162 (2024) 56-68   https://doi.org/10.21741/9781644903018-3

Chapter 3

Thermoelectric Properties of Polymer and Organic-Inorganic Composites

Sahar Sarfaraz[1], Maria Wasim[1]*, Aneela Sabir[1], Muhammad Shafiq[1]

[1]Institute of polymer and textile engineering, Univeristy of the Punjab, Lahore, Pakistan

maria-be24@hotmail.co.uk, aneela.ipte@pu.edu.pk

Abstract

This chapter provides insights of the thermoelectric properties of polymer in addition to organic and inorganic composites. Devoid of any maintenance requirement or moving parts, thermoelectric energy generation offers incredibly reliable, small sized, light and quiet operative power sources with no moving parts and no maintenance requirements. About two centuries ago, thermoelectricity was used as an alternate of petroleum power plants.

Keywords

Thermoelectric Polymer, Organic, Inorganic, Composites

Contents

1. Introduction

Materials which directly convert heat to electricity are called thermoelectric materials. To maximize the efficiency of global energy, electricity is derived from wasted heat. Approximately around two-third of the energy consumed by the industries and private households is lost to the environment due to resource contraints. As the heat quality is poor therefore can not be employed. Shortcomings of low quality waste heat recovery procedures are making usable energy harvesting operations expensive. Numerous natural resources like solar, wind and geothermal energy are giving great potentials for energy consumption. The unrelenting depletion of oil and gas drives a strong demand for green energy technologies as well as its conventionally produced, cost-effective uses.

With no moving components and no maintenance needs, thermoelectric energy generation offers exceptionally dependable, small sized, light, and quiet operating power sources. For two centuries, thermoelectricity is being used as a substitute for fossil fuel-powered power plant.

Excellent thermoelectric properties revealed by the inorganic semiconductors like Bi_2Te_3, $PbTe$, $SiGe$, Cu_2Se and $SnSe$ but their applications have been limited because of high price and processing difficulty [1-3].

2. Thermoelectric polymers

Polymeric materials are viewed as potential choice because of low weight, mechanical flexibility and characteristic low thermal conductivity [4, 5]. Conducting conjugated polymers are most common thermoelectric polymer materials that display adjustable electrical conductivity over a wide range at doping level e.g., polyacetylene (PA) [6], polythiophene (PTh) [7], poly (3,4-ethylenedioxythiophene) (PEDOT) [8] and polyaniline (PANI) [9] and polypyrrole (PPy) [10]. Their structures are presented in Fig. 1:

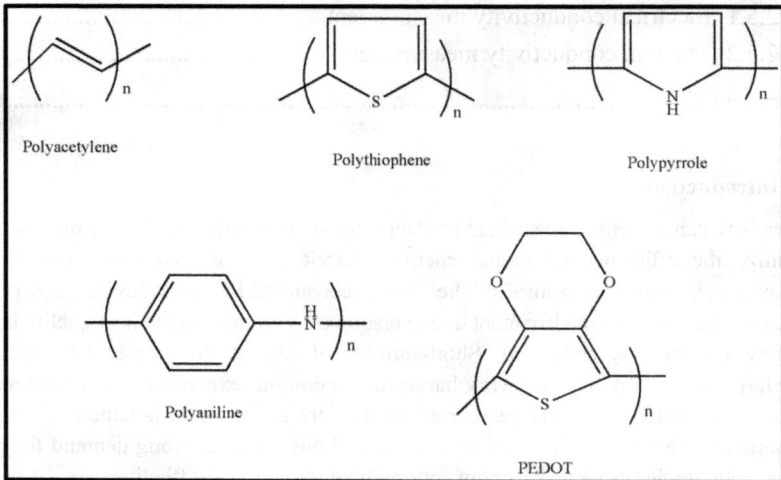

Fig. 1: Structures of some conducting polymers

2.1   Thermoelectric organic-inorganic composites

Polymer/inorganic composites have high Seebeck coefficient values, their electrical conductivity values cannot satisfy the needs. The principle challenge is to shape the close interfacial interactions among polymer and inorganic particles because of the absence of effective associations [11]. Due to which carrier movement is restricted in polymer/inorganic composites. For instance, around 100 S cm^{-1} electrical conductivity value is shown by the composites of PEDOT:PSS blended with p-type Bi_2Te_3 powders which was lower than the values of both PEDOT:PSS and Bi_2Te_3 [12]. This low value emerged from the huge interfacial resistance among both substances.

Low thermal conductivity values are also observed because of poor connections among polymer and inorganic particles and low electrical conductivity. Investigation of in-plane thermal conductivity of PEDOT:PSS/SnSe nanocomposite films was done using LFA 447 nanoflash apparatus [13]. While increasing the loading of SnSe nanosheets from 0 to 50 wt%, the thermal conductivity values can also be expanded from 0.25 to 0.45 $Wm^{-1} K^{-1}$. These reduced thermal conductivity values were attributed to the strong phonon dispersing at the interfaces in polymer/inorganic composites, in this manner advancing the accomplishment of outstanding thermoelectric execution.

2.2   Thermoelectric properties

Merit (a dimensionless figure) is closely related to transformation efficiency of thermoelectrics [14] that is defined as in eq 1:

$$ZT = \frac{S^2 \sigma T}{k} \tag{1}$$

Where, S = Seebeck co-efficient, σ = Electrical conductivity, k = Thermal conductivity and T = Absolute temperature

So, larger value of merit is directly proportional to electrical conductivity and seebeck co-efficient and shows an inverse relation with thermal conductivity. Power factor is proportional to electrical conductivity:

$$PF = S^2 \sigma \tag{2}$$

Prevailing movement of charge carriers (electrons and holes) is associated with electrical conductivity of TE material while thermal conductivity is related to photons displacments. Electrical conductivity can be written as in eq 3: [15]

$$\sigma = n\mu e \tag{3}$$

Where, n = Concentration of charge, e = Electron charge and μ = Mobility of charge

From equation 3, it can be assumed that for better thermoelectric performance of material, quicker portability and higher concentration charge carriers are helpful. Conducting polymers charge carriers concentration is impacted by the doping level and their portability is connected to polymer structure and morphology [16]. Therefore electrical conductivity and Seebeck co-efficient are interlinked. By lowering the concentration of charge carrier, it can build up the Seebeck co-efficient while declining the electrical conductivity. Thus, for obtaining high power factors, there is an ideal doping level. N-type and p-type TE polymers show a great impact as shown in Fig 2 [16].

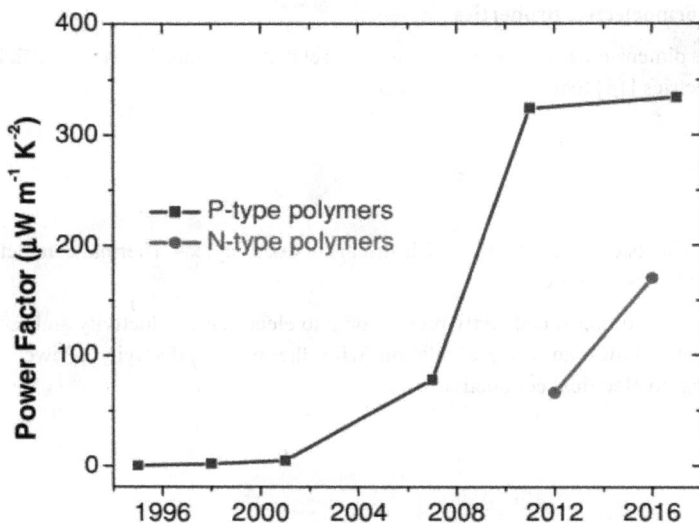

Fig 2: N-type and P-type TE polymers power factors [16]

Another well known methodology is to frame composites of TE polymers with nanomaterials. The nanofillers added in polymer matrix can build up the electrical conductivity or Seebeck co-efficient of the polymer lattice, while thermal conductivity will not increase at their lowest loading. Subsequently, TE polymer composites can have low thermal conductivity and high mechanical flexibility like the TE polymers. Fillers are of two types:

Inorganic particles like Te nanofillers [17] SnSe [18] and Bi2Te3 [19] giving larger S (Seebeck co-efficient) value.

Carbon particles graphene [20], carbon nanotubes (CNT) [21] having high electrical conductivity.

2.3   Thermoelectric effects

Any phenomena in which there is replacement of heat and electrical energy is called thermoelectric effect. A thermodynamically reversible process is that in which last state of a system is reestablished similar to its initial state where there is no creation of entropy or energy dissipation. Three reversible thermoelectric effects occur in these processes:

Thermoelectric Polymers: Properties and Applications Materials Research Forum LLC
Materials Research Foundations 162 (2024) 56-68 https://doi.org/10.21741/9781644903018-3

- Seebeck Effect.... Used for thermoelectric production.

- Peltier effect....... Used for electric refrigeration.

- Thomsan Effect...No practical use.

Simultaneously, thermoelectric gadget's performance remains lower than the Carnot proficiency because of two present irreversible cycles that incorporate Joule heating and thermal conduction. Since, it is essentially difficult to isolate reversible and irreversible cycles in thermoelectrics, it must be dealt with non-equilibrium thermodynamics.

2.3.1 Seebeck effect

German physicist Thomas Johann Seebeck developed a relationship between electricity and heat from his experiments. He used an open circuit comprising of two different materials (standard thermocouple design) in his experiment as shown in figure 3 where cold and hot junctions were kept at temperature T1 and T2 respectively making ΔT as temperature gradient. In these conditions, a circuit was generated with an electromotive force (Seebeck voltage). Here, V and T are directly related via a proportionate coefficient known as the thermopower or seebeck coefficient [22]. Charge carriers that are available for conduction in a material diffuse from hotter region to cooler portion maintained at lower temperature as a result of temperature gradient effect. After this process has continued, a steady Seebeck voltage formed as a force acting against further movement of charge transporters, resulting in a constant state. For holes, the Seebeck coefficient is positive, while for electrons, it is negative. Naturally, V depends on the type of material (n- or p-type) and the temperature gradient.

So the expression of a thermocouple Seebeck coefficient is given in Eq 4:

$$S_A - S_B = \frac{\Delta V}{\Delta T} \tag{4}$$

Basically the Seebeck coefficient of a single homogeneous conductor cannot be calculated directly because the resulting electromotive force is zero. So a mixture of two distinct materials is always used for estimating Seebeck coefficient values.

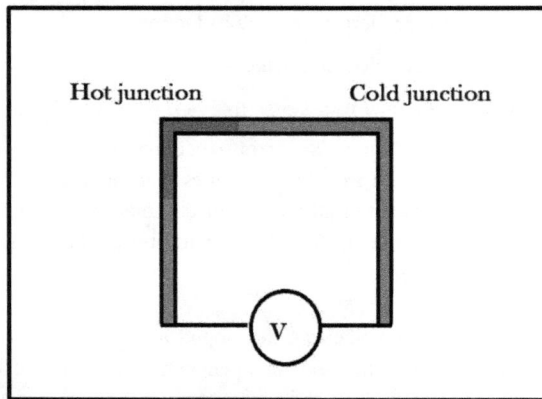

Fig 3: Demonstration of Seebeck effect in a circuit including two different materials

2.3.2 Peltier effect

A French watchmaker Jean Peltier discovered the Peltier effect in 1834. He used similar experiment as that of Seebeck effect assimilation where no temperature gradient was laid out between the intersections. In this case, current moving through the coil comprised of two conductors creating cooling and heating at the juncture [23] Fig 4.

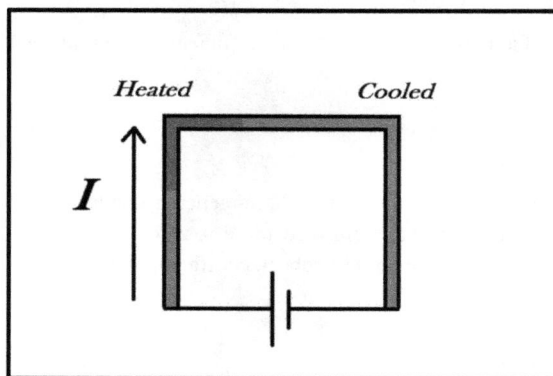

Fig 4: Circuit of two different materials representing Peltier effect

Electric current was proportional to the heat absorption and release rate with proportionality coefficient known as Peltier coefficient [24]. At contant T, Peltier coefficient for single material is given as in eq 5:

$$\Pi = Q/I \tag{5}$$

Heat absorbed at junction for circuit with two distinct materials A and B is:

$$Q = \Pi_{AB} I = (\Pi_A - \Pi_B)I \tag{6}$$

I stands for electric current, and is the Peltier coefficient at the junction of two conductors. Positive Peltier coefficient is shown by p-type material while n-type substance gives negative value. In an n-type substance, electrons move from the negative to positive end delivering energy in the form of heat during electric field existence. While in a p-type material, holes in the form of heat move to the negative end inverse of current stream. Thus, switching the electric flow polarity and redirecting the heat flow.

2.3.3 Thomson effect

A mix of above two effects called Thomson effect which depicts the heat production when an electric flow moves through an inconsistently heated individual conductor [25]. (Fig 5)

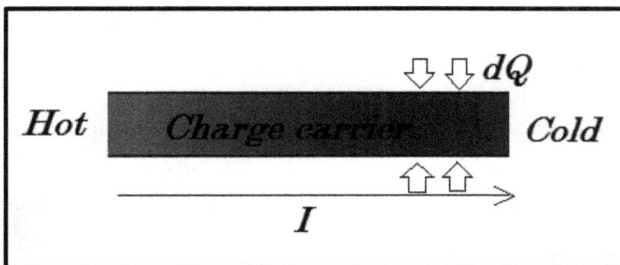

Fig 5: Single conductor Thomson effect

The substance warming/cooling rate is in direct relation to temperature gradient and electrical flow with a Thomson coefficient as proportionately coefficient.

$$Q\,Thomson \;=\; \tau\,I\,(-\Delta T) \tag{7}$$

The sign of TC indicates either the electric current streams from cooler to hotter side or vice versa in a conductor. TC and SC are connected by first Kelvin relation:

$$\tau \;=\; T\,dS/dT \tag{8}$$

Reference to the type of the substance, Thomson coefficient can be positive or negative and if S is temperature autonomous then τ is zero. There is no significance of thomson effect in the activity of TE gadgets while it cannot be ignored in experiments requiring great precision.

2.4 Joule heating and thermal conduction

Alongside the three reversible TE effects, two irreversible processes joule warming and thermal conduction bring down the TE gadget's performance [26]. Gadget's productivity is blocked from achieving its thermodynamic limits due to the occurrence of these effects called Carnot proficiency.

Flow of electric current results in joule warming while in thermoelectric case, it is unwanted as it changes valuable electric energy into heat which dissipates and called Ohmic loss. Because of thermal conduction, the temperature of cold end increases in a conductor which can be expressed as:

$$Q \;=\; -Ak\frac{dT}{dx} \tag{9}$$

A = Cross section area of a conductor   **K** = Thermal conductivity

Decrease in TE effectiveness occurring due to thermal conduction is in direct relation with K thus for better energy transformation, less thermally conductive substances are essential.

2.5 Measurement techniques

2.5.1 Electrical conductivity measurement

Four point probe process is used to measure voltage (Fig 6). Four evenly spaced silicon or glass coated electrodes are connected to the thin film of TE material through one set of lead.

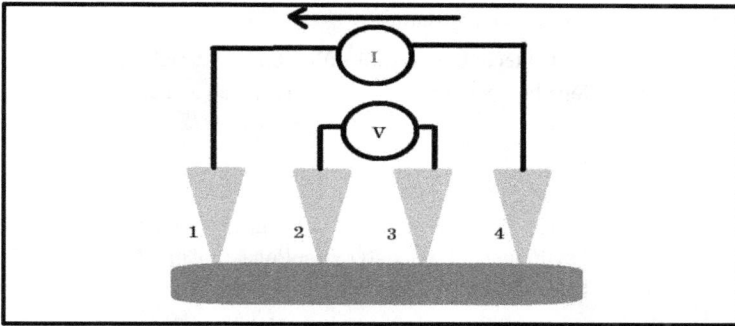

Fig 6: Four point probe method to measure electrical conductivity

Seebeck Coefficient Measurement:

Calculation of Seebeck coefficient can be done at the juncture made of two different conductors from temperature and voltage gradient measurement [27].

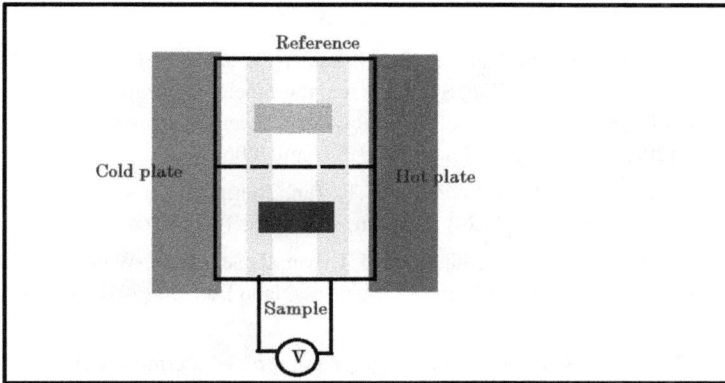

Fig 7: Seebeck coefficient measurement on thin film

2.5.2 Thermal conductivity measurement

It is measured by using 3ω technique for thin film, bulk materials and liquids [28]. 3ω voltage gives information about the amount of heat dissipated and can be used to calculate heat conductivity [28]

References

[1] F. Jiang, X.J. Kun, B. Lu, X. Yu, H.R. Jin, L.L. Feng, Thermoelectric performance of poly(3,4-ethylenedioxythiophene): poly(styrenesulfonate), Chin. Phys. Lett. 25 (2008) 2202-2205. https://doi.org/10.1088/0256-307X/25/6/076

[2] Y.W. Park, W.K. Han, C.H. Choi, H. Shirakawa, Metallic nature of heavily doped polyacetylene derivatives: Thermopower, Phys. Rev. B. 30 (1984) 5847-5851. https://doi.org/10.1103/PhysRevB.30.5847

[3] H. Yao, Z. Fan, H. Cheng, X. Guan, C. Wang, K. Sun, J. Ouyang, Recent development of thermoelectric polymers and composites, Macromol. Rapid Commun. 39 (2018) 1700727. https://doi.org/10.1002/marc.201700727

[4] D.S. Maddison, J. Unsworth, R.B. Roberts, Electrical conductivity and thermoelectric power of polypyrrole with different doping levels, Syn. Metal. 26 (1988) 99-108. https://doi.org/10.1016/0379-6779(88)90339-6

[5] I. Lévesque, P.-O. Bertrand, N. Blouin, M. Leclerc, S. Zecchin, G. Zotti, C. Ratcliffe, D. Klug, X. Gao, F. Gao, J. Tse, Synthesis and thermoelectric properties of polycarbazole, polyindolocarbazole, and polydiindolocarbazole derivatives, Chem. Mater. 19 (2007) 2128. https://doi.org/10.1021/cm070063h

[6] M. Lepinoy, P. Limelette, B. Schmaltz, F.T. Van, Thermopower scaling in conducting polymers, Sci. Reports. 10 (2020) 8086. https://doi.org/10.1038/s41598-020-64951-z

[7] K. See, J. Feser, C. Chen, A. Majumdar, J. Urban, R. Segalman, Water-processable polymer-nanocrystal hybrids for thermoelectrics, Nano Lett. 10 (2010) 4664-4667. https://doi.org/10.1021/nl102880k

[8] D. Kim, Y.S. Kim, K. Choi, J. Grunlan, C. Yu, Improved thermoelectric behavior of nanotube-filled polymer composites with poly(3,4-Ethylenedioxythiophene): poly(Styrenesulfonate), ACS Nano. 4 (2010) 513-523. https://doi.org/10.1021/nn9013577

[9] H. Wang, C. Yu, Organic thermoelectrics: Materials preparation, performance optimization and device integration, Joule. 3 (2019) 53-80. https://doi.org/10.1016/j.joule.2018.10.012

[10] J. Mao, G. Chen, Z. Ren, Thermoelectric cooling materials, Nat. Mater. 20 (2021) 454-461. https://doi.org/10.1038/s41563-020-00852-w

[11] H. Xi, L. Luo, G. Fraisse, Development and applications of solar-based thermoelectric technologies, Renew. Sustain. Ener. Rev. 11 (2007) 923-936. https://doi.org/10.1016/j.rser.2005.06.008

[12] M. Lu, X. Zhang, J. Ji, X. Xu, Y. Zhang, Research progress on power battery cooling technology for electric vehicles, J. Energ. Storage. 27 (2020) 101155. https://doi.org/10.1016/j.est.2019.101155

[13] Y. Gurevich, J.V. Pérez, Peltier Effect in Semiconductors, 2014. https://doi.org/10.1002/047134608X.W8206

[14] B. Sherman, R. Heikes, R. Ure, Calculation of efficiency of thermoelectric devices, J. Appl. Phys. 31 (1960) 1-16. https://doi.org/10.1063/1.1735380

[15] M.J. Klein, P.H.E. Meijer, Principle of minimum entropy production, Phys. Rev. 96 (1954) 250-255. https://doi.org/10.1103/PhysRev.96.250

[16] Y. Demirel, Fundamentals of non-equilibrium thermodynamics, in: Y. Demirel (Eds.), Non-equilibrium Thermodynamics, Elsevier, Amsterdam, 2014, pp. 119-176. https://doi.org/10.1016/B978-0-444-59557-7.00003-5

[17] C.A. Domenicali, Irreversible thermodynamics of thermoelectricity, Rev. Modern Phys. 26 (1954) 237-275. https://doi.org/10.1103/RevModPhys.26.237

[18] J. Callaway, Model for lattice thermal conductivity at low temperatures, Phys. Rev. 113 (1959) 1046-1051. https://doi.org/10.1103/PhysRev.113.1046

[19] G.J. Snyder, E.S. Toberer, Complex thermoelectric materials, Nature Mater. 7 (2008) 105-114. https://doi.org/10.1038/nmat2090

[20] M. Wasim, A. Sabir, R.U. Khan, Membranes with tunable graphene morphology prepared via Stöber method for high rejection of azo dyes, J. Envtal. Chem. Eng. 9 (2021) 106069. https://doi.org/10.1016/j.jece.2021.106069

[21] M. Wasim, S. Sagar, A. Sabir, M. Shafiq, T. Jamil, Decoration of open pore network in polyvinylidene fluoride/MWCNTs with chitosan for the removal of reactive orange 16 dye, Carbohydr. Polym. 174 (2017) 474-483. https://doi.org/10.1016/j.carbpol.2017.06.086

[22] B. Poudel, Q. Hao, Y. Ma, Y. Lan, A. Minnich, B. Yu, X. Yan, D. Wang, A. Muto, D. Vashaee, X. Chen, J. Liu, M. Dresselhaus, G. Chen, Z. Ren, High-thermoelectric performance of nanostructured bismuth antimony telluride bulk alloys, Science. 320 (2008) 634-638. https://doi.org/10.1126/science.1156446

[23] Z. Lee, Y. Tang, D. Zou, Thermoelectric and stress distributions around a smooth cavity in thermoelectric material, Inter. J. Mech. Sci. 107198 (2022). https://doi.org/10.1016/j.ijmecsci.2022.107198

[24] W.M. Yim, F.D. Rosi, Compound tellurides and their alloys for peltier cooling-A review, Solid-State Electron. 1972. 15 (1972) 1121-1140. https://doi.org/10.1016/0038-1101(72)90172-4

[25] V. Kuznetsov, L.A. Kuznetsova, A.E. Kaliazin, D. Rowe, Preparation and thermoelectric properties of A8IIB16IIIB30IV clathrate compounds, J. Appl. Phys. 87 (2007) 7871-7875. https://doi.org/10.1063/1.373469

[26] K.L. Jablonska, Semiconductors, 2022.

[27] W. Luo, H. Li, Y. Yan, Z. Lin, X. Tang, Q. Zhang, C. Uher, Rapid synthesis of high thermoelectric performance higher manganese silicide with in-situ formed nano-phase of MnSi, Intermetallics. 19 (2011) 404-408. https://doi.org/10.1016/j.intermet.2010.11.008

[28] M. Shikano, R. Funahashi, Electrical and thermal properties of single-crystalline (Ca2CoO3)0.7CoO2 with a Ca3Co4O9 structure, Appl. Phys. Lett. 82 (2003) 1851-1853. https://doi.org/10.1063/1.1562337

Thermoelectric Polymers: Properties and Applications
Materials Research Foundations 162 (2024) 69-80

Materials Research Forum LLC
https://doi.org/10.21741/9781644903018-4

Chapter 4

Materials used in Thermoelectric Polymers

Muhammad Khalil[1,] Sana Kainat[1], Maria Wasim[1]*, Aneela Sabir[1], Muhammad Shafiq[1]

[1] Institute of Polymer and textile engineering, University of the Punjab, Lahore, Pakistan

maria-be24@hotmail.co.uk, aneela.ipte@pu.edu.pk

Abstract

Nowadays, Polymer based thermoelectric (TE) materials are gaining more and more attention from researchers because of their easy processability, low thermal conductivity, and low cost. These polymers contain conjugation in their structure which helps to conduct electricity through them. Common thermoelectric materials contain conducting polymers and inorganic dopants. The resulting thermoelectric material may be of n-type or p-type. Their properties depend upon the morphology of matrix material, the concentration of charge carriers, and dopants... In this chapter, different types of conducting polymers (CE) and their derivatives, synthesis, properties, and factors affecting the properties of CE will be discussed.

Keywords

Thermoelectric Polymers, Conducting Polymers

Contents

1.   Introduction

Due to rapid growth in population and industries, the demand of energy has increased [1] which results in the rise in prices and depletion of existing energy resources. This problem forced the researchers and scientists to explore new areas to meet the global demand for cheap and continuous energy. Thermoelectric materials which convert heat energy into useful energy, have drawn the attention of researchers. The term used for such materials is thermoelectric generator (TEGs). These materials are being widely used in military, medical and electrical industries [2]. Based on their chemical composition, thermoelectric materials include organic, inorganic, polymer, organic-inorganic hybrid and metal-organic coordination polymers. Organic TE materials gained attention because of their flexibility, low thermal conductivity and light weight. The efficiency of energy conversion of TEGs is calculated by figure of merit (ZT) which is given in equation (1).

$$ZT = \frac{S^2 \sigma T}{k} \tag{1}$$

In equation (1) the S stands for Seebeck coefficient, σ stands for electrical conductivity of material, K stands for thermal conductivity and absolute temperature is represented by T.

[3]. The figure of merit is gained with low thermal conductivity, high electrical conductivity, and high Seebeck coefficient. The equation (2) gives the value of power factor (PF). It is proportional to the value of ZT.

$$PF = S^2 \sigma \tag{2}$$

Many inorganic and organic TEs are being used, all over the world, in different electrical equipment's. Also, the performance of TEs can be improved by improving their ZT value. The ZT value of inorganic TEs has been higher [4]. The toxic elements in inorganic TEs e.g. Sb, Pb, Te, and Bi are substituted with nontoxic elements like, Se and Sn. As compared to inorganic TEs, organic TEs are, light in weight, and have low thermal conductivity i.e. ranged from 0.1 to 1 $Wm^{-1} k^{-1}$. Organic thermoelectric polymers contain conjugated system in their backbone which enable them to conduct electricity. Polyanilines, polythiophenes, polypyrroles, and polyacetylenes are the organic polymers utilized in thermoelectrics. In order to increase the electrical conductivity of polymers, inorganic nanoparticles are added to other TEs, which are polymer-based composites. For instance, iodine-doped polyacetylene films are strongly conductive and have a power factor of 105 Scm^{-1}, which is sufficient for thermoelectric applications. These polymers are synthesized through doping and solution based processes. Thermoelectric materials consisting of polymer have vast and applications in heat conversion foils and sensors. In this chapter different types of polymer based TEs will be discussed.

2.   Conducting polymers

Polymers are made up of carbon based repeating units. Some polymers are singly bonded while others have conjugation in their chain structure. The conjugated polymers have ability to conduct electrical charges which originates through π-conjugation system. A typical thermoelectric polymer is made up of several monomers that are conjugated to one another along the length of the chain. Due to their short chain lengths, conjugated polymers are stiff during chain bending and twisting. Additionally, they are not particularly soluble, and hence some side groups are added to improve their solubility. The structures of some conjugated polymer is shown in Figure 1.

In conjugated polymers the charge is carried through polaron (carrying single charge) or bipolarons (carrying radical ion pair formed by the removal of electron from polaron). It is suggested that these charge carriers are created upon doping of polymers. Unlike doping in inorganic material, doping in conjugated polymers is done through chemical doping technique in which oxidizing or reducing agents are used. In this process electron transfer between polymers and dopants takes place. In electrochemical doping, a cell is constructed between conjugated polymer and electrode, which provides charge to the polymer. The electrolyte between electrodes maintain the neutrality of the polymer by ion transfer. Most of the doped polymers are prepared through this process. Other methods of doping include Photo doping [6], in which photovoltaic devices are used, based on absorption of radiation and charge separation, doping by treatment with acids or bases [7] and doping by charge injection without counter ion involvement (OFET) [8].

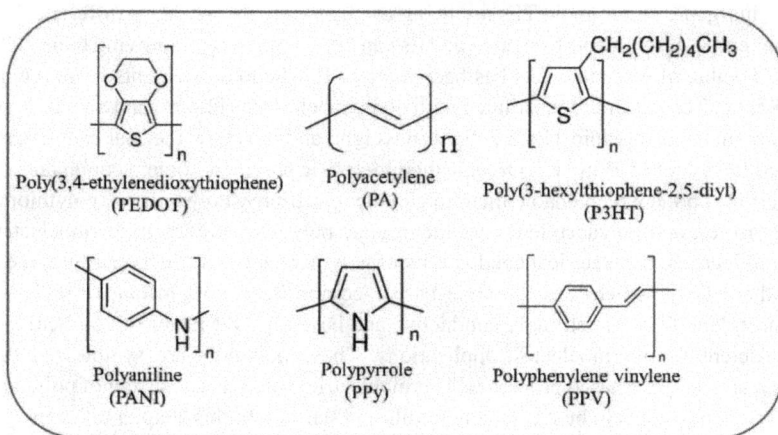

Figure 1: Structure of some conjugated polymers [5].

Morphology of polymers also plays a major role in carrier transport in thermoelectric materials. The ordered structure of conjugated polymers cannot be maintained up to larger lengths. Although, polymers have some amount of crystallinity [9], yet they mostly present in amorphous form. Despite the higher concentration of carriers, amorphous form does not allow the charge mobility. So, for higher charge mobility, optimum ordering in structure is required.

2.1 Preparation and processing of thermoelectric polymers

Thermoelectrics with desired properties (flexibility, electrical and mechanical) can be attained by mixing inorganic materials with polymers. The synthesis of inorganic polymer composites render superior performance as compared to the individual components. The polymer composites have been prepared through solution mixing, in-situ preparation and physical mixing.

In solution mixing, the polymer solution contains dispersed inorganic fillers. It is then followed by evaporation which, sometimes, results in homogeneous dispersion of inorganic fillers [10]. In the physical mixing method, inorganic components are mixed and dispersed in the melt of polymer, inorganic fillers are dispersed on the basis of liquid particle dispersion and extruded to get desired properties of inorganic polymer composites.

Materials Research Foundations 162 (2024) 69-80   https://doi.org/10.21741/9781644903018-4

In, in-situ preparation method, inorganic fillers are mixed with monomer solution before the polymerization. The demerit of this technique is that the inorganic filler may separate or sediment from organic polymer solution because of weak Vander walls forces among polymers and fillers. The properties of conducting polymers are enhanced by different techniques like doping, post treatments. Some common conductive polymers and their thermoelectric properties are discussed below:

3. P-type thermoelectric polymers

The polymeric materials show high conducting, thermal and environmental stability, low band energy gap, strong backbone, and lightweight. The following are some P-type thermoelectric conducting polymers [11].

3.1 Polyacetylene

Polyacetylene doped with iodine vapor shows high electrical conductivity. The temperature can influence the electrical conductivity of material and with the doping of $FeCl_3$ polyacetylene stretch films shows maximum conductivity at 220 K. The poor air stability and insolubility limit the use of polyacetylene [12].

3.2 Polyaniline

The most attractive and useful functional polymers are conductive polymers. Iodine doped polyacetylene was the first conductive organic polymer with vast applications in secondary batteries, light emitting diodes due to mechanical properties of electronic and inorganic semiconductors. Polyaniline molecules consist of two repeating units, reduced unit and oxidized unit. Polyaniline behaves differently when the ratios of these units change in polyaniline. It acts as an insulator in fully reduced and oxidized form but when the ratio of both units is about 0.5, it shows maximum conductivity [13].

Organic thermoelectric polymers based on polyanilines have a wide range of applications due to their outstanding manufacturing, preparation, and cost-effectiveness characteristics. The thermoelectric figure of merit (ZT) was utilised to assess the properties. Toshima *et al.* [13] prepared multilayered films of polyaniline which improve performance through quantum effect. The polyaniline shows the same ZT effect as FeSi2. The polyaniline exhibits various structures on which its electric properties depend (Figure 2).

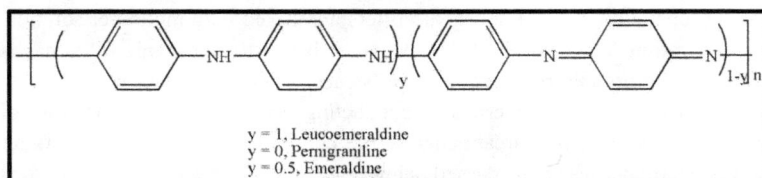

y = 1, Leucoemeraldine
y = 0, Pernigraniline
y = 0.5, Emeraldine

Figure2: Polyaniline structure at different oxidation states [13].

3.3   Polypyrrole

It exhibits low cost, easy processing and excellent environmental stability and is widely used in energy devices. Nanocomposites based on polypyrrole have been the most commonly used. The carbon nanotubes were most commonly used in doping of thermoelectric materials by in-situ polymerization or dissolution in organic solvent which contain multiwall carbon nanotube suspensions. Its nanostructure effects the properties of material and used in different forms like nanoparticles, nanotubes and nanowires [12].

3.4   (3,4-ethylenedioxythiophene)

It is known as (PEDOT) and used in a wide range of applications including sensors, photovoltaics, thin film transistors and light emitting diodes. In oxidation state, it forms stable and transparent films and exhibits low redox potential while in neutral state it is insoluble. To overcome solubility problem, polystyrene sulfonate is used. The PSS based PEDOT has low conductivity which is improved through doping of various polar solvent (Figure 2). The PEDOT-based tosylate exhibits excellent conductivity and environmental stability and the Seebeck coefficient [5, 14].

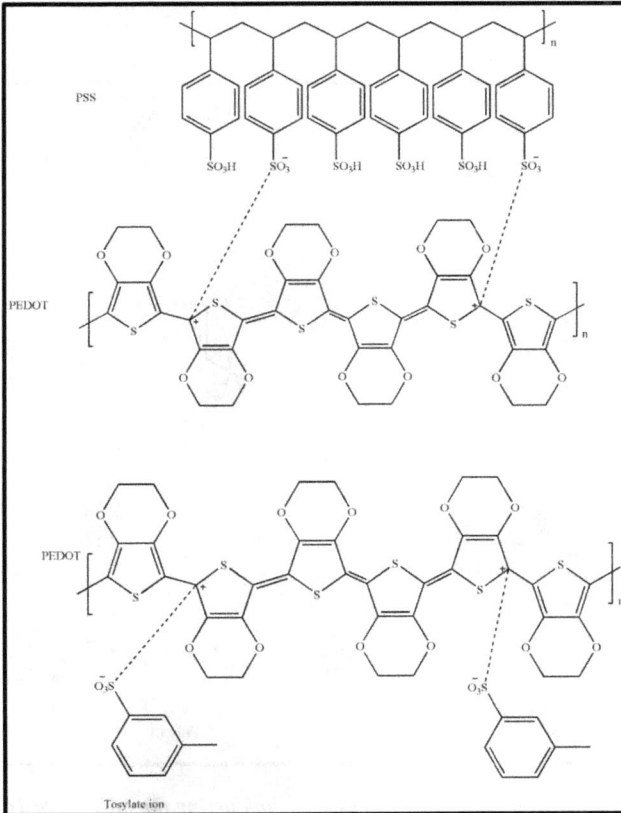

Figure 3: Chemical structure of Polystyrene sulfonate or tosylate based PEDOT [5].

3.5 Polythiophenes

A high Seebeck coefficient, improved tensile strength, low thermal conductivity, and high electrical conductivity are all characteristics of poly(3-methylthiophene) nanofilms. ZT value is displayed as 3.0×10^{-2}. It shows excellent performance due to better arrangement of polymer chains and orientation in films. Its polymer films were doped with $NOPF_6$ at 220 to 370 K. The polymer was very sensitive to the doping procedure. The counter ions of PF_6^- in an environment electron can move. When the electrical conductivity increased significantly, the Seebeck coefficient fell. The poly(alkylthiphene) blend polymer was prepared and has ground state carrier. The ZT and absolute conductivity was not particularly high [15].

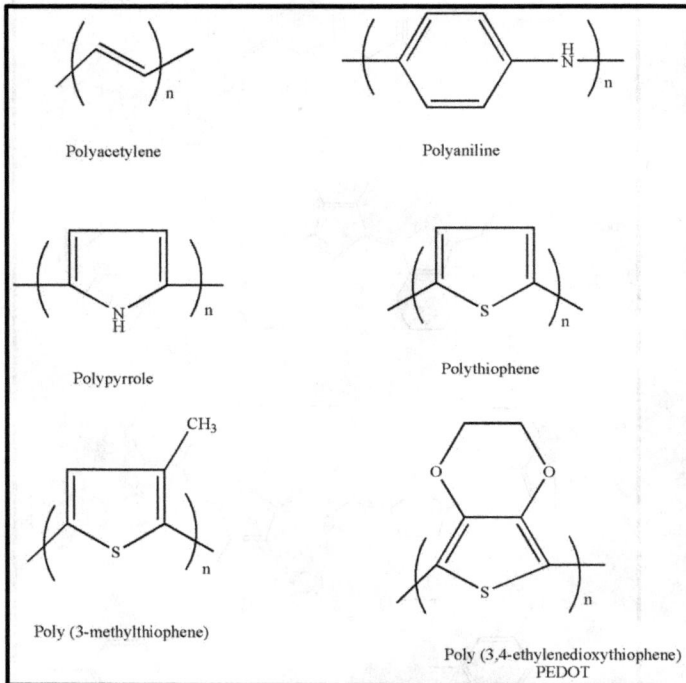

Figure 4: Some important un-doped conducting polymers [15, 16].

3.6   Poly(2,7-carbazole) and derivative

When this main chain was introduced, the carbazole unit forms linkage at 3,6-position and in presence of pendant group 9-position and it shows easy substitution at this position. It shows good potential for thermoelectrically applications. The nitrogen atom present in this chain was oxidized prior to carbon backbone and its charge shows localization. Due to its structure a large Seeback coefficient was obtained. Due to negative charge carrier pinning it has a negative effect on polymer conductivity. The poly(2,7-carbazolylenevinylene) derivative was used to overcome this drawback. The vinylene unit enhances the conductivity of polymer without lowering Seeback coefficient. The vinylene unit in 2,7-carbazole based polymer was replaced with electron-donating conjugated units to increase the stability and electric conductivity. The nature of the side chain attached with nitrogen atom was significant to improve molecular organization and solubility of thin films. The doping increases the thermal properties and electrical conductivity [15].

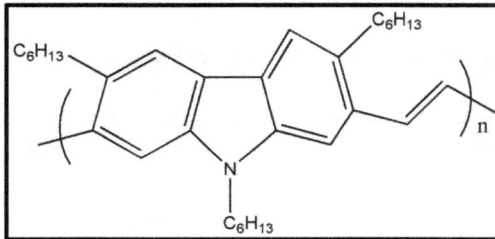

Figure 5: Structure of Poly(2,7-carbazolylenevinylene) [15]

4.   n-type thermoelectric polymers

It shows poor thermal conductivity if compared with the p-type thermoelectric polymers due to incomplete doping. n-type TE is air sensitive and its stability was related to the LUMO level. If LUMO level depth increase, then its conductivity increased. When carbon nanotubes mixed with the conjugated polymers, it results in long lifetime and air stability [15].

The most commonly used n-type polymers include perylene bisimide, benzotizaole and fullerenes. Its LUMO level of building units can be arranged through different electron donating units [15].

4.1   Factors affecting thermoelectric properties

The conducting polymers properties were determined through chemical and structural design. Following factors affect the properties of thermoelectric materials [17, 18].

4.1.1 Polymer structure

The outcome of the structure will help in understanding of thermal, and electrical conductivities as well as figure of merits. Bond conjugation was the molecular basis for determining the properties of conductivity [17].

4.1.2 Concentration of polymer

During synthesis, polymer concentration helps to determine the molecular weight/chain length which affects the viscosity and solubility. Different sizes of nanostructures resulting out through different molecular weights, affect the charge transportation. Electrical conductivity and the Seebeck coefficient are affected by charge carrier density [17].

4.1.3 Temperature

The temperature plays a major role in affecting the charge carrier density as well as photon-electron scattering. The non-metallic behavior was displayed through electrical conductivity and at low doping the temperature increased. The superposition was increased through high conversion efficiency [17].

4.1.4 Polymer chain alignment

The electrical conductivity was increased through the alignment of polymer chains. The stretching of CP results in an increase in conductivity in stretched direction. The aligned chain increases the carrier mobility and neutralizes the positive charge. The alignment of the polymer backbone also helps in increase of power factor and electrical conductivity [17].

References

[1] Y.J. Zeng, D. Wu, X.H. Cao, W.X. Zhou, L.M. Tang, K.Q. Chen, Nanoscale organic thermoelectric materials: Measurement, theoretical models, and optimization strategies, Adv. Funcn. Mater. 30 (2020) 1903873. https://doi.org/10.1002/adfm.201903873

[2] M. Dargusch, W.D. Liu, Z.G. Chen, Thermoelectric generators: Alternative power supply for wearable electrocardiographic systems, Adv. Sci. 7 (2020) 2001362. https://doi.org/10.1002/advs.202001362

[3] R. Kroon, D.A. Mengistie, D. Kiefer, J. Hynynen, J.D. Ryan, L. Yu, C. Müller, Thermoelectric plastics: From design to synthesis, processing and structure-property relationships, Chem. Soc. Rev. 45 (2016) 6147-6164. https://doi.org/10.1039/C6CS00149A

[4] C. Di, W. Xu, D. Zhu, Organic thermoelectrics for green energy, Natl. Sci. Rev. 3 (2016) 269-271. https://doi.org/10.1093/nsr/nww040

[5] J.L. Bredas, S.R. Marder, The WSPC Reference on Organic Electronics: Organic Semiconductors, 2016, World Scientific, p. 277-298. https://doi.org/10.1142/9678-vol1

[6] D. Nilsson, M. Chen, T. Kugler, T. Remonen, M. Armgarth, M. Berggren, Bi-stable and dynamic current modulation in electrochemical organic transistors, Adv. Mater. 14 (2020) 51-54. https://doi.org/10.1002/1521-4095(20020104)14:1<51::AID-ADMA51>3.0.CO;2-#

[7] J.M. Zhuo, L.H. Zhao, R.Q. Png, L.Y. Wong, P.J. Chia, J.C. Tang, S. Sivaramakrishnan, M. Zhou, E.C.W. Ou, S.J. Chua, Direct spectroscopic evidence for a photodoping mechanism in polythiophene and poly (bithiophene-alt-thienothiophene) organic semiconductor thin films involving oxygen and sorbed moisture, Adv. Mater. 21 (2009) 4747-4752. https://doi.org/10.1002/adma.200901120

[8] B. Hamadani, D. Corley, J.W. Ciszek, J. Tour, D. Natelson, Controlling charge injection in organic field-effect transistors using self-assembled monolayers, Nano Lett. 6 (2006) 1303-1306. https://doi.org/10.1021/nl060731i

[9] A.B. Kaiser, Systematic conductivity behavior in conducting polymers: Effects of heterogeneous disorder, Adv. Mater. 13 (2001) 927-941. https://doi.org/10.1002/1521-4095(200107)13:12/13<927::AID-ADMA927>3.0.CO;2-B

[10] D.X. Crispin, Retracted article: Towards polymer-based organic thermoelectric generators, Ener. & Env. Sci. 5 (2012) 9345-9362. https://doi.org/10.1039/c2ee22777k

[11] B.T. McGrail, A. Sehirlioglu, E. Pentzer, Polymer composites for thermoelectric applications, Angew. Chem. Int. Ed. 54 (2015) 1710-1723. https://doi.org/10.1002/anie.201408431

[12] C.J. Yao, H.-L. Zhang, Q. Zhang, Recent progress in thermoelectric materials based on conjugated polymers, Polymers. 11 (2019) 107. https://doi.org/10.3390/polym11010107

[13] H. Yan, N. Sada, N. Toshima, Thermal transporting properties of electrically conductive polyaniline films as organic thermoelectric materials, J. Therm. Anal. Calorim. 69 (2002) 881-887. https://doi.org/10.1023/A:1020612123826

[14] X. Guan, E. Yildirim, Z. Fan, W. Lu, B. Li, K. Zeng, S.-W. Yang, J. Ouyang, Thermoelectric polymer films with a significantly high Seebeck coefficient and thermoelectric power factor obtained through surface energy filtering, J. Mater. Chem. A. 8 (2020) 13600-13609. https://doi.org/10.1039/D0TA05324D

[15] N. Dubey, M. Leclerc, Conducting polymers: Efficient thermoelectric materials, J. Polym. Sci. Part B: Polym. Phys. 49 (2011) 467-475. https://doi.org/10.1002/polb.22206

[16] Y. Du, S.Z. Shen, K. Cai, P.S. Casey, Research progress on polymer-inorganic thermoelectric nanocomposite materials, Prog. Polym. Sci. 37 (2012) 820-841. https://doi.org/10.1016/j.progpolymsci.2011.11.003

[17] M. Bharti, A. Singh, S. Samanta, D. Aswal, Conductive polymers for thermoelectric power generation, Prog. Mater Sci. 93 (2018) 270-310. https://doi.org/10.1016/j.pmatsci.2017.09.004

[18] M.A. Kamarudin, S.R. Sahamir, R.S. Datta, B.D. Long, M.F.M. Sabri, S.M. Said, A review on the fabrication of polymer-based thermoelectric materials and fabrication methods, Sci. World J. 713640 (2013) 1-17. https://doi.org/10.1155/2013/713640

Thermoelectric Polymers: Properties and Applications
Materials Research Foundations 162 (2024) 81-98

Materials Research Forum LLC
https://doi.org/10.21741/9781644903018-5

Chapter 5

Cage Structured Compounds

Anita Gupta[1*], H. Kaur[2], Ishika Aggarwal[1], Koushiki Chatterjee[1]

[1] Amity Institute of Applied Sciences, Amity University, Noida (U.P.), India

[2] Applied Sciences, Punjab Engineering College (Deemed to be University), Chandigarh, India

agupta3@amity.edu

Abstract

This chapter has targeted the features of inorganic cage compounds, their classification, and their overabundant applications. The term 'Cage' is utilized as a specific term for three-dimensional structures which have a definite and rigid geometry. There can be various atomic positions in the cage, these positions can operate as branching origins and additional ligands can be introduced at these positions. Recent advances in the synthetic chemistry and strategies adopted for the synthesis of cage compounds with special attention to Calixarenes and Cryptophanes have played an analytical role in the development of Biomedical Applications, drug freightage systems, sensing, bio-imaging, and other smart materials. The adsorption of dye from industrial wastewater by the functionalized Calixarene cage e.g. Dinitro Calix [4] arene cage (DNCC) was tested for the first time [1]. Recent research indicates that Cryptophane cages can also adopt different configurations and demonstrate various applications [2]. During the 1970s, a founding work by Corbett and Simon on rare-earth halides led to the establishment of the fact that these highly cluster skeleton electron-deficient clusters always involve an interstitial atom inside the metal cluster cage. A novel development in the field of coordination chemistry of integral cage molecules and their ligand complexes has been extensively researched for their budding application as building blocks in polymers [3]. Caged structure complexes, one of its kind in supramolecular chemistry have always garnered the attention of the scientific community worldwide. This book chapter is an initiative to instill the essence of the caged compound with its striking features and a plethora of applications.

Keywords

Cage Compounds, Calixarenes, Cryptophanes, Metal Cluster Cage, Supramolecular Chemistry

Contents

1.   Introduction

Caged complexes, as the name suggests, are a consequence of host-visitant complexation, through various types of interactions such as electrostatic, hydrophobic, hydrogen bonding, van der Waals, and π-π bonding. With the changing times, the nature of scientific exploration has changed, but our host receptacle has stood the test of time. The classification has not changed ever since its inception.

Although, it is said the more we see, the more we learn and the more we learn, we evolve to serve better to society. But in this case, our previous generations had discovered the host molecules and the chemistry behind them. And now we, the future generations are set to explore its applications. Self-arrangement through the mediation of metal has produced structural motifs as in helicases, grids, links, knots, spheres, and cages, with cages drawing the most attention due to their nanoscale holes. Application as a focused container and constrained response environment is made possible by molecular orchestration acting as hosts. The arena has recently advanced significantly thanks to the implementation of specialized functionality, such as catalytic centers or photoswitches that provide stimulation control.

To ease the classification, caged structure complexes are classified based on the following different parameters:

- Classification based on the mode of synthesis.

- Classification on the basis of various fields of applications.

- Classification based on their mechanism of complexation.

Let's discuss these classifications in detail:

2.   Classification based on the mode of synthesis

- Caged ATP and cAMP were among the first caged compounds that living cells could synthesize. The ternary treads of caged ATP were created by connecting NPE-caged phosphate to adenosine diphosphate; however, caged cAMP was built by the hyper-reactive diazo reaction which allowed opinionated confinement of phosphates [4].

- "Coordination cages", termed as "Metal–Organic Polyhedrons", or also called as "Metal–Organic Super Containers abbreviated as MOCS" very similar to "Porous Coordination Cages abbreviated as PCCs", stand as distinct supramolecular existents comprising metal interface and organic moieties [5].

- MOFs are drawing increasing attention due to the structural modifications, having large specific surface area, better porosity, and predictable pore space framework [6]. Owing to their structural variegation, porous framework, customizable orifice size, and thermal and chemical stability, it has garnered attention globally [7]. These supramolecules are crafted from organic chelates and metal frameworks [8, 9]. A general scheme of multi-step and single-step synthesis is illustrated in scheme 1.

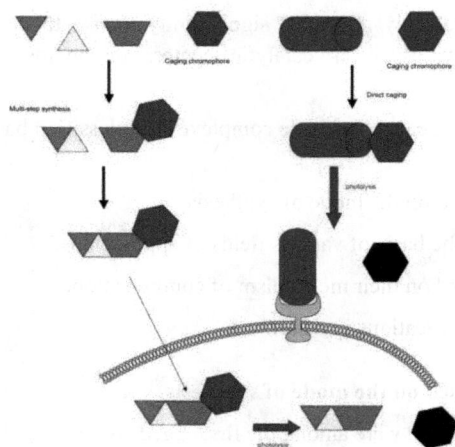

Scheme 1: General scheme for single step and multi-step caging.

Classification on the basis of various fields of applications:

3. Biomedical Applications

- Major progress in the field of material chemistry has allowed the merger of novel nanoparticles with customized optical and chemical attributes with better efficiency [10]. MOCs have portrayed themselves as a wonder material in the area of biochemistry, underpinning its marvelous host–visitant activity. Platinum or palladium-containing MOCs has shown eminent performance as an anticancer agent with higher selectivity and prominent IC50 value when contrasted with cisplatin and DOX [11-14].

- While as drug freightage systems, MOCs depicted remarkable binding ability and regulated discharge of visitant molecules under external stimulation [15]. They have been investigated as potential anti-cancer drugs against fifteen distinct forms of cancer, including malignant tumours in the lungs, breasts, ovaries, liver, prostate, stomach, skin, mouth, thyroid, and other malignancies. In this application, MOCs containing platinum, palladium, and ruthenium are primarily employed [16].

- MOCs having multiple receptacles have been explored as drug freightage and are prepped for cramming and discharge of various drugs.

Recognition of enantiomers and their separation: Gas Chromatography, chiral fluorescence, and potentiometric sensor technique, and enantio-selective adsorption have been in use for more than a decade, but chiral molecule-based cages have shown remarkable results in this field aided by GC [17].

Anion identification: Entrapment of anions leads to prismatic and fluorescence alteration, thereby allowing for visitant detection at a minute concentration [18]. The revered supramolecule has been investigated in various fields such as catalysis, metal sequestration, pollution separation and carbon dioxide conversion [19].

Chromatographic solid substrate: MOCs have been successfully explored for its role as sorbent materials in extraction strategies and as stationary phases in chromatographic techniques [20]. The formation of a novel homochiral pentyl cage molecule involved the condensation of (S,S)-1,2-pentyl-1,2-diaminoethane and 1,3,5-triformylbenzene. It was further diluted with polysiloxane (OV-1701) and studied as a novel stationary phase for high-resolution gas chromatographic separation of organic compounds. Baseline separation of several positional isomers occurred on the capillary column coated with a pentyl cage [21].

Targeted Freightage: With potential uses in guest delivery, gas storage, separation, heterogeneous catalysis, or luminescence, MOCs have demonstrated photo-thermally induced resonance [22].

Cage-gel smart materials were created with their programmable features, and they have molecule separation, catalysis, and luminous materials as potential uses for them [23].

Fullerenes are carbon footballs that have attracted global attention for a wide range of applications. Vacant cage fullerenes have remarkable electrochemical attributes with a potentially favorable biotic property. A different variety of fullerene can accommodate metals inside them. The framework of fullerene has an enclosure type of architecture with very economical reorganization energy low-lying excited states and elongated triplet lifetimes [24].

Therapeutic application: Genetically based accommodation of therapeutic moieties into the pharmacologically active metal enclosure. When this orchestration is done with protein-based freightage, it has displayed higher therapeutic benefit when contrasted with free drugs in both in vivo and in vitro. Accommodation of multifunctional groups which can be customized to the peripheral surface and can mediate effectively at the target site [25].

4.   Classification based on their mechanism of complexation

Reversibly nonreactive, reversibly reactive, and irreversibly reactive are the three general analyte identification routes. As they bind and release an interest ion without a chemical reaction, small molecule establishing elements like crown ethers exhibit reversibly nonreactive recognition properties. The best-known revocable reactive recognition elements are boronic acids, which react reversibly with glucose and other saccharides. A reversibly reactive system may still contain enzymes; one example is glucose sensors that use glucose oxidase. Whether the recognition element and reporter are irrevocably altered or reversibly linked is crucial. Local pH changes can result in the coupling of the ionophore to a reporter, which then generates an ion indication [27].

Special attention has been paid to Cryptophane cages and Calixarene cages due to their vast applications and extensive continuing research in these fields. The synthesis of these solid-state functionalized materials with inestimable large structures using different metal ions and organic spacer ligands has been an exponentially rising pitch of research in the past decade. An overview of these cage compounds is discussed here.

4.1   Cryptophane cages

Cryptophanes comprise three-fold cyclotribenzylene (CTB) units and have a diadem architecture, therefore it has a receptacle to house visitants of choice [28]. Owing to their stereochemistry, cryptophanes can be customized by the nature of the colligates connecting CTBs. Global attention on cryptophanes is mostly due to their capacity to engulf and interact with visitant in a selective fashion. Usually, visitant complexation depends upon the dimensions of the receptacle and the configuration of the visitant. Prudent placement of moieties on the peripheral surface or interior of the receptacle has the ability to after binding strengths, solubility, and other attributes. Targeted imaging and biological identification of avidin protein by employing xenon-functionalized cryptophane with a biotin functional group have been deployed at various places. This was further coupled with HyperCEST method to achieve the threshold for effective sensing [29].

In another separate study, MS_2 viral capsid linked to cryptophane had achieved a LOD of 0.7pM. Likewise, molecular cancer cell imaging was accomplished by linking bacteriophage to 330 cryptophane, such that it could cohere with EGFR cell surface receptors (generally found in solid tumors due to retro cell coding leading to overexpression of the cancer gene) [30].

In a separate study, metalloproteinase – 7, an enzyme generated in large quantities in various cancers was detected by the grasping of cryptophane. The analysis was carried by investigating the blood work of the subject [30].

In another scientific exploration, to identify mercuric ions, a biosensor was designed by linking two cryptophanes with a dipyrrolylquinoxaline group. On interaction with the analyte a clasp-like orchestration was formed which resulted in a chemical shift [129]Xe NMR (nuclear magnetic resonance) as the electron clouds of cryptophanes overlay one another. This effect also facilitated to detect [129]Xe MRI (magnetic resonance imaging) contrast in the residence of mercuric ions, when compared to the blank biosensor [29]. A visual representation in given in scheme 2.

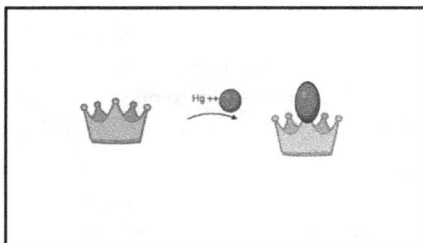

Scheme 2: Cryptophane exhibiting a contrast in [129] Xe MRI in the company of mercuric ions.

In a similar fashion, bio thiols were identified using acrylate-based cryptophane, the contrast was detected when incubated with cysteine. Separate research findings demonstrated that on coupling cryptophane A and 4-azide-1,8-naphthalic anhydride, a biosensor was created which could detect HS⁻ by the change in [129]Xe NMR (nuclear magnetic resonance) chemical shift after the reduction of azide group to amine group by induction of HS⁻ [31]. An interpretation of the change is given in Scheme 3.

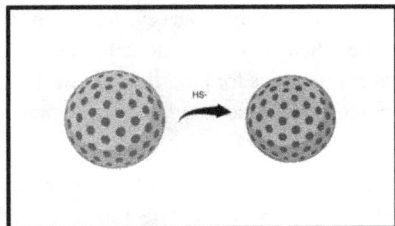

Scheme 3: Fluorescence was observed only in the presence of HS⁻.

4.2   Calixarene cages

Calixarenes have been in the limelight for their unique features and applications from their inception as a side product of bakelite formation via condensation of phenol and formaldehyde. It constitutes as a subset of cyclo-oligomers. Their structural robustness comes from their basket-like shape with a well – defined top and bottom rims along with a central annulus. Hence, these features bestow them with the power to act as a receptacle for foreign material [32].

Just like crown ethers and cyclodextrins, calixarenes act as prime receptacle molecules. It's simple to prepare calixarenes by the condensation reaction of aldehydes with phenols having a remarkable yield. It was discovered in the late eighteenth century by Adolph von Bayer. At the end of the late nineteen seventies, the reaction of methanal and hydroxybenzene had evolved and was named "Calixarene" [33, 34].

Due to its unique shape and reactive sites in calixarene, it has the potential to be utilized for varied functions. The supramolecular structure has a non-polar edge, a polar edge, and an aquaphobic container. The sides of the receptacle may be modified specifically to support analyte-selectivity or to facilitate chemical change. Its inner cavity has the ability to host completely different visitant molecules; thus, calixarene-based polymers fabricate a chance for fine-tuning the scaffold specifically in analytical chemistry [35-37].

Generally, calixarene-based macromolecules are synthesized through chain reaction having monomers having calixarene or immobilizing of calixarene monomers on a polymeric cast. It provides a chance to create materials having extraordinary attributes. The polymeric molecules containing calixarene units on the rim/in substituted lateral fetters possess the potential to be utilized in the form of sensors, biosensors, selective membranes, catalysts, macromolecules, and analyte-specific freightage [38-43]. Calixarene-centered amphiphilic copolymer orchestration in aqueous medium was examined in contemporary investigation [44, 45]. During the literature survey, Researchers from all over the world have been interested in the chemistry of producing polymers with a calix[n]arene framework and its potential applications for bipartite subsets involving "covalent polymers made by supramolecular containing monomers" and " polymers which are functionalized via calixarene moieties."

This receptacle has an exceptional capacity to play host to a plethora of visitant molecules, based on their morphology, which is easily fine–tuned through the dimensions of an addendum as a component for the framework. Amid the above-mentioned macrocyclic structures, calixarenes have drawn the attention of the scientific community for their utilization in the analyte identification of harmful substances. Attributes such as non – pernicious nature, facile preparation, the existence of chemically instinctive sites, 3-

dimensional cavities which can selectively form complexes with guest molecules or ions are several attributes that make these molecules pertinent in the research community.

Fine-tuning of calixarenes has been under scrutiny for a long time and has enabled researchers to drastically change the attributes of parent macrocycles. As already mentioned, the easiness rendered to the upper edge or lower edge for functionalization with moieties like azo, amides, imines, sulfur, semicarbazone, and alkyl groups to list a few. This attribute permits the existence of a varied range of supramolecular compounds having various features of identification, specificity, solubility, and degree of aquaphobicity.

Calixarenes can be functionalized straight from the starting materials, exempli gratia, with different natures of substituents on hydroxybenzene, the upper edge alteration can be facilitated.

The Locus of reactivity is fixed at two regions: Lower edge hydroxyl groups or upper-edge alcohol groups. Atypical functionalities like carboxylates, phosphates, ammonium groups, or sulfonates, are fixed into the lower rostrum via facile reactions with available precursors and can be specifically prepared with astonishing yields.

Alterations are normally done via:

- Upper rostrum – functionalization at the fourth position of the aromatic rings.
- Lower rostrum – functionalization of alcoholic moieties of hydroxybenzene.

4.3   Upper rostrum alteration

Alteration of the upper rostrum of calixarenes includes the replacement of moiety at the fourth position to the hydroxybenzoic oxygen atom. Aluminum trichloride removal of tertiary butyl moiety can be done in the company of phenol or toluene. And post departure of the alkane moiety, electrophilic replacement can be done on the fourth position. Replacement at this particular location covers halogenation, nitration, sulphonation, sulfo-chlorination, acylation, chloro-methylation, and amino methylation. Once these substituents have been introduced, they undertake further reactions like reduction of the nitro groups or aryl-aryl Suzuki coupling to name a few examples. The para-substituent can be easily removed by $AlCl_3$ -catalyzed trans-alkylation.

After the deletion of tert- butyl groups from calixarene, it is ready for functionalization [46].

4.4   Lower rostrum alteration

This rostrum is less pressurized for alteration. Although the potential utilization of the lower rostrum bears more prospects than the higher rostrum alterations. Depending on the extension of the orifice, the lower rostrum has the ability for preparing complexes with higher-grade groups such as toxic metals. Scientific exploration till now has implied the potential of morphological alteration of calixarenes, which include facile engulfment of drugs, aromatic compounds and toxic metals. Therefore, the alcoholic groups of hydroxybenzene at lower rostrum presents itself as an remarkable reactive site for origination of moieties which alter the configuration and complexation attributes of the above mentioned compounds [47].

Esters of calixarenes were the earliest derivatives to be prepared. When bases weaker than NaH are present, it is frequently possible to selectively produce partly substituted calixarenes from acid halides by utilising some esterifying reagent or other bulky esterifying reagents. An early example for selective functionalization is p-tert-butyl calix [4] arene reaction with benzoyl chloride in the presence of pyridine to give the tribenzoate [48-50].

5-. Polymers designed by covalent bonding of monomers having calixarene moiety

The synthesis of calixarene containing macromolecule follow two-fold strategies. First, is based upon the polymerization of a functionalized calixarene moiety, and the other one supports fixture of the calixarene unit onto a polymer matrix.

6. Calixarene functionalized polymers used for iodine capture

Micronutrients are vital nourishment components required by our body, one of them is iodine. It plays a key role in thyroid gland in generating and processing hormones. The accumulation of iodine in the ecosystem is considered as an associate in environmental degradation that has major impact on health and ecology. The micro essential halide is additionally a noteworthy radionuclide that has severe impact on the public health at large. Hence, it is crucial to detect and eliminate iodine from water. Till now, many polymers bearing calixarene backbone have been synthesized. These studies have revealed that various kinds of adsorbents have been tested to remove extra I_2 from water and ecosystem.

For instance, calix[4]arene-derived 2-dimensional macromolecule was obtained following the cross-coupling process with a calix with a bromo functional[4] Diethynyl-1,10-biphenyl and arene tetrol which is mixed in anhydrous 1,4-dioxane (Sonogashira–Hagihara). Its characterization indicated the polymer was fluffy and segregated as a few-layer thick nanosheet. When tested as an iodine vapor adsorbent, the data was supported with excellent efficiency. Its regeneration via washing by ethanol was also reported with a really little loss of potency [39, 51]. Its diagrammatic representation is given in Scheme 4.

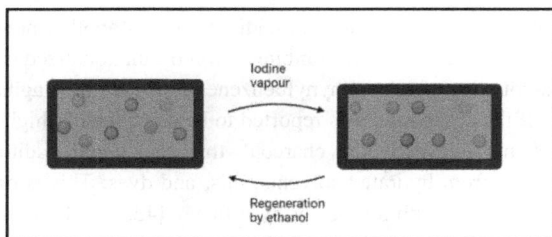

Scheme 4: Calix nano sheet acting like a nano sponge to adsorb iodine vapour and further regenerated with the help of ethanol.

In a separate finding organic polymers were tested for pollutant sponges. They demonstrated their ability to adsorb iodine nearly double the mass in iodine vapours and additionally, acted as a good adsorbent for Congo red dye from water-based solution having a pH range of 2 – 10. The anchorage of congo red on to the polymer is illustrated in scheme 5.

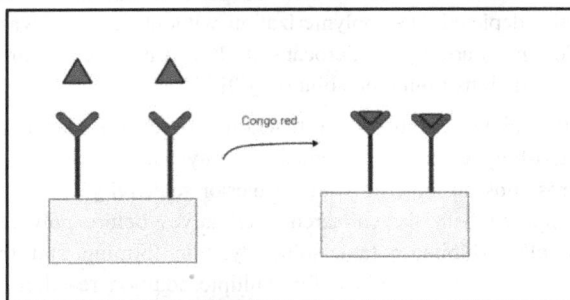

Scheme 5: Viologens based calixarene used for entrapment of congo red.

7.   Sensing and elimination of pollutants.

One of the most pertinent obligations of industrial work is expelling hepatotoxic miscible and non-biodegradable pollutants from effluents. Various biotic and chemical strategies have been made accessible for expelling adulterants present in effluents and among them, sorption is known to be the best process. A reputed category of artificial receptor molecules is calixarene. Their characteristic receptacle cavity can accommodate neutral, ionic analytes guests by substantiated, selective, and specific interlinkage of vivid genesis.

The preparation of new polymers using amido calix[4]resorcinarene unit as a macrocyclic adsorbent towards different radical organic dyes like congo red, methyl orange etc. (all of which are miscible in aqueous medium). In addition, to this another novel calix[4]arene-funtinalized polymer was successively combined bromo functionalized calix[4]arene with Pd catalyzed cross-coupling of 1,4-diethynylbenzene (Sonogashira–Hagihara). The surface adhering capacity of the compound was reported to be considerably higher in comparison to different adsorbents like activated charcoal, this also could additionally eliminate organic contamination from hydrated solvents, oils, and dyes. This is often used for the sorption/desorption process with a noteworthy efficacy [42, 51]. In another study, to get rid of heavy metals like lead, cadmium, cobalt, mercury and arsenic (in their ionic form) from water refuse by means of extraction (solid–liquid) using polycalix[4]arene and polycalix[4]amide consisting of a moiety which is mesogenic triazole are used [52].

A calixarene-based microcapsule was employed in an undisturbed solution for polymerization. This microcapsule manifested exceptional withdrawing potential for basic radicals. Microcapsule was examined through FTIR (Fourier transformed infrared spectroscopy technique) and XPS (X-ray photoelectron spectroscopic technique) analysis. Moreover, TEM (Transmission electron microscopic analysis) and DLS (Dynamic light scattering analysis) depicted That polymerization without agitation was found to be a superior methodology to arrange a microcapsule. It will be able to extract a significant amount of heavy metal ions from their solution [40].

It was possible to achieve a condensation reaction between 1,3-adamantane, p-tert-butyl calix[4]arene, p-tert-butyl calix[8]arene, and p-tert-butyl calix[4]resorcinarene. As a result, 3-dimensional cross-linking material with a superior selectivity for few metal ions was reported and compared with the calixarene derivatives before polymerization. These polymers additionally exhibited notable solubility, film–forming, and, thermal stability. Furthermore, they were also examined for multiple addition reactions of 1,6- hexane diisocyanate and p-test-butyl calix[8]arene in triethylamine that resulted in another soluble polymer in high yield [49, 53].

As reported in a separate study, a series of novel optically active polymers containing calix[4]arene moieties were prepared via multiple condensation reactions with calix[4] arene-based compounds. Solid–liquid sorption was employed to know the complexing ability of polymers with specific alkali metals and hepatotoxic transition metals in different oxidation states. Conclusive outcome was there, when compared to the calix[4]arene precursor. It depicted that the synthesized polymers have dependable ionophores for the detection of basic radicals like sodium & potassium and, additionally for hepatotoxic significant metal cations like copper, cobalt, cadmium, and mercury [54, 55].

Calixcrowns belong to the calixarenes clan which exhibits prearranged framework and robust attachment sites in comparison to calixarenes and crown ethers. Till this time, various findings have been reported on calixcrowns. Perse, calixcrown polymers were reported and synthesized by condensing calix[6]arene hexa-esters. Research on the surface assimilation attributes of those polymers with some cations has also been proclaimed. The surface assimilation results for sodium, potassium, silver, mercury, copper, cobalt, and nickel ions were appreciable. Its pictorial representation is presented in Scheme 6.

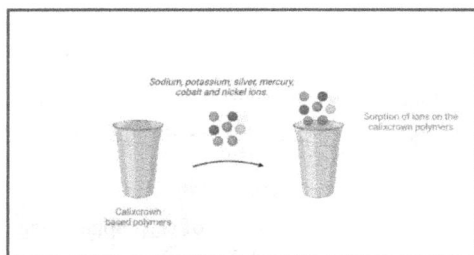

Scheme 6: Sorption of sodium, potassium, silver, mercury, copper, cobalt, and nickel ions on Calix-based polymers.

Calix[6]hydroxy amide-based composite was developed to entrap uranium, thorium, and cerium in the ionic form To ingrain on the matrix of non-aberrant polymers such as dextrans and for metal complexation, the multistep procedure for a unique disubstituted calix[4]arene was created. At the last leg of the reaction, the orchestrated medallion was entrenched to the matrix employing a cyanuric linker. The synthesized blend expressed a fascinating capacity for complexation in the company of cuprous ions [56].

Normally, the measure for standard of life is by the grade of water used or consumed, which additionally determines the well-being of the general public. An efficient functionalized material was insinuated (by the interaction of calix[4]arene derivative and XAD-4) for the isolation and recovery of chromium(VI) from an aqueous medium. Various criteria such as, modification in pH scale, concentration, and height of the floor were improved and also the elevated surface assimilation capabilities were recorded at 85mg/g. Additionally, at optimal surroundings, this particular material was recovered to 97% [39].

Conclusion & future challenges

This book chapter is aimed at introduction, classification, applications, and recent advancements of cage structure complexes in varied fields. A special mention of cryptophanes and calixarenes has been made to draw attention to the recent scientific exploration in this field. Despite various potentials offered by the above-mentioned supramolecules, there is still room for improvement and optimization of its applications in biomedical and sensing arenas.

There is significant potential to exploit the characteristic attributes of supramolecules for various additional applications in biochemistry and imaging. There are still a large number of diversified molecules where auxiliary complexes are yet to be prepared and their attributes are hitherto explored.

References

[1] F. Temel, M. Turkyilmaz, S. Kucukcongar, Removal of methylene blue from aqueous solutions by silica gel supported calix[4]arene cage: Investigation of adsorption properties, Eur. Polym. J. 125 (2020) 109540. https://doi.org/10.1016/j.eurpolymj.2020.109540

[2] 0.D. Negra, Access to the Syn diastereomers of cryptophane cages using HFIP, Chem. Commun. 58 (2022) 3330-3333. https://doi.org/10.1039/D1CC06607B

[3] J. Wachter, Coordination Polymers with Group 15/16 Element Building Blocks, in: J. Reedijk, K. Poeppelmeier (Eds.), Comprehensive Inorganic Chemistry II, Elsevier: Amsterdam, 2013, p. 933-952. https://doi.org/10.1016/B978-0-08-097774-4.00139-X

[4] K. Svoboda, R. Yasuda, Principles of two-photon excitation microscopy and its applications to neuroscience, Neuron. 50 (2006) 823-839. https://doi.org/10.1016/j.neuron.2006.05.019

[5] Y. Fang, Catalytic reactions within the cavity of coordination cages, Chem. Soc. Rev. 48 (2019) 4707-4730. https://doi.org/10.1039/C9CS00091G

[6] J. Liu, M. Chen, H. Cui, Recent progress in environmental applications of metal-organic frameworks, Water Sci. Technol. 83 (2021) 26-38. https://doi.org/10.2166/wst.2020.572

[7] N. Manousi, Extraction of metal ions with metal-organic frameworks, Molecules, 24 (2019) 4605. https://doi.org/10.3390/molecules24244605

[8] M. Shen, Antibacterial applications of metal-organic frameworks and their composites, Compr. Rev. Food Sci. Food Saf. 19 (2020) 1397-1419. https://doi.org/10.1111/1541-4337.12515

[9] E.-S.M. El-Sayed, D. Yuan, Metal-organic cages (MOCs): From discrete to cage-based extended architectures, Chem. Lett. 49 (2020) 28-53. https://doi.org/10.1246/cl.190731

[10] D. Zhang, Metal-organic cages for molecular separations. Nat. Rev. Chemi. 5 (2021) 168-182. https://doi.org/10.1038/s41570-020-00246-1

[11] A. Tarzia, K.E. Jelfs, Unlocking the computational design of metal-organic cages, Chem. Commun. 58 (2022) 3717-3730. https://doi.org/10.1039/D2CC00532H

[12] E. Raee, Y. Yang, T. Liu, Supramolecular structures based on metal-organic cages, Giant 5 (2021) 100050. https://doi.org/10.1016/j.giant.2021.100050

[13] G.C.E. Davies, Caged compounds: Photorelease technology for control of cellular chemistry and physiology, Nat. Methods 4 (2007) 619-628. https://doi.org/10.1038/nmeth1072

[14] A. Katoch, N. Goyal, S. Gautam, Applications and advances in coordination cages: Metal-Organic Frameworks, Vacuum 167 (2019) 287-300. https://doi.org/10.1016/j.vacuum.2019.03.038

[15] C.Y. Zhu, M. Pan, C.Y. Su, Metal-organic cages for biomedical applications, Isr. J. Chem. 59 (2019) 209-219. https://doi.org/10.1002/ijch.201800147

[16] D.T. Walters, Utilization of a nonemissive triphosphine ligand to construct a luminescent gold (I)-box that undergoes mechanochromic collapse into a helical complex, J. Am. Chem. Soc. 140 (2018) 7533-7542. https://doi.org/10.1021/jacs.8b01666

[17] J.H. Zhang, Recent advances of application of porous molecular cages for enantioselective recognition and separation, J. Separ. Sci. 43 (2020) 134-149. https://doi.org/10.1002/jssc.201900762

[18] P.P. Neelakandan, A. Jiménez, J.R. Nitschke, Fluorophore incorporation allows nanomolar guest sensing and white-light emission in M 4 L 6 cage complexes, Chem. Sci. 5 (2014) 908-915. https://doi.org/10.1039/C3SC53172D

[19] A. Giri, Cavitand and molecular cage-based porous organic polymers, ACS Omega 5 (2020) 28413-28424. https://doi.org/10.1021/acsomega.0c04248

[20] P.R. Bautista, Metal-organic frameworks in green analytical chemistry, Separations, 6 (2019) 33. https://doi.org/10.3390/separations6030033

[21] S. Xie, J. Zhang, N. Fu, B. Wang, Application of homochiral alkylated organic cages as chiral stationary phases for molecular separations by capillary gas chromatography, Molecules 21 (2016) 1466. https://doi.org/10.3390/molecules21111466

[22] L. Feng, K.Y. Wang, G.S. Day, The chemistry of multi-component and hierarchical framework compounds, Chem. Soc. Rev. 48 (2019) 4823-4853. https://doi.org/10.1039/C9CS00250B

[23] I. Jahović, Cages meet gels: Smart materials with dual porosity, Matter, 4 (2021) 2123-2140. https://doi.org/10.1016/j.matt.2021.04.018

[24] A.D. Kelkar, D.J. Herr, J.G. Ryan, Nanoscience and nanoengineering: Advances and applications, CRC Press, 2014. https://doi.org/10.1201/b16957

[25] E.J. Lee, N.K. Lee, I.-S. Kim, Bioengineered protein-based nanocage for drug delivery, Adv. Drug Deliv. Rev. 106 (2016) 157-171. https://doi.org/10.1016/j.addr.2016.03.002

[26] R.J. Drout, L. Robison, O.K. Farha, Catalytic applications of enzymes encapsulated in metal-organic frameworks, Coord. Chem. Rev. 381 (2019) 151-160. https://doi.org/10.1016/j.ccr.2018.11.009

[27] H. Shet, A comprehensive review of caged phosphines: Synthesis, catalytic applications, and future perspectives, Organic Chem. Front. 8 (2021). 1599-1656. https://doi.org/10.1039/D0QO01194K

[28] T. Traoré, Scalable synthesis of cryptophane-1.1. 1 and its functionalization, Org. Lett. 12 (2010) 960-962. https://doi.org/10.1021/ol902952h

[29] Y. Voloshin, I. Belaya, R. Krämer, Cage metal complexes: Clathrochelates revisited, 2017, Springer. https://doi.org/10.1007/978-3-319-56420-3

[30] A. Casini, B. Woods, M. Wenzel, The promise of self-assembled 3D supramolecular coordination complexes for biomedical applications, ACS Publications, 2017, p. 14715-14729. https://doi.org/10.1021/acs.inorgchem.7b02599

[31] A.K. Gupta, Mapping the Assembly of Metal-Organic Cages into Complex Coordination Networks, 2017.

[32] A. Dondoni, Synthesis and characterization of calix[4]arene-based copolyethers and polyurethanes Ionophoric properties and extraction abilities towards metal cations of polymeric calix[4]arene urethanes, Polymer 45 (2004) 6195-6206. https://doi.org/10.1016/j.polymer.2004.06.012

[33] C.D. Gutsche, Topics in calixarene chemistry, J. Incl. Phenom. Macrocycl. Chem. 7 (1989) 61-72. https://doi.org/10.1007/BF01112783

[34] C.D. Gutsche, B. Dhawan, K. No, Calixarenes. 4. The synthesis, characterization, and properties of the calixarenes from p-tert-butylphenol, J. Am. Chem. Soc. 103 (1981) 3782-3792. https://doi.org/10.1021/ja00403a028

[35] U. Balami, D.K. Taylor, Electrochemical responsive arrays of sulfonatocalixarene groups prepared by free radical polymerization, Reactive Funct. Polym. 81 (2014) 54-60. https://doi.org/10.1016/j.reactfunctpolym.2014.03.015

[36] A.R. Mendes, Linear and crosslinked copolymers of p-tert-butylcalix[4]arene derivatives and styrene: New synthetic approaches to polymer-bound calix[4]arenes.

Reactive Funct. Polym. 65 (2005) 9-21.
https://doi.org/10.1016/j.reactfunctpolym.2005.01.006

[37] Y. Yang, T.M. Swager, Main-chain calix[4]arene elastomers by ring-opening metathesis polymerization, Macromolecules 40 (2007) 7437-7440. https://doi.org/10.1021/ma071304+

[38] T.C. Gokoglan, A novel architecture based on a conducting polymer and calixarene derivative: Its synthesis and biosensor construction, RSC Adv. 5 (2015) 35940-35947. https://doi.org/10.1039/C5RA03933A

[39] Y. Zhang, Robust cationic calix[4]arene polymer as an efficient catalyst for cycloaddition of epoxies with CO2, Indus. Eng. Chem. Res. 59 (2019) 7247-7254. https://doi.org/10.1021/acs.iecr.9b05312

[40] Z. Zhu, D. Liu, Q. Ren, Y. Tan, Microcapsule dispersion of poly (calix[4]arene-piperazine) for hazardous metal cations removal from waste water, Iranian Polym. J. 28 (2019) 697-706. https://doi.org/10.1007/s13726-019-00739-x

[41] T. Nekrasova, Structural and dynamic characteristics of a star-shaped calixarene-containing polymer in aqueous solutions: The formation of mixed-shell micelles in the presence of poly (methacrylic acid), Polym. Sci. Series A 57 (2015) 6-12. https://doi.org/10.1134/S0965545X15010071

[42] A. Ten'kovstev, A. Razina, M. Dudkina, Synthesis and complexing behavior of amphiphilic starlike calix[4]arenes, Polym. Sci. Series B 56 (2014) 274-281. https://doi.org/10.1134/S1560090414030166

[43] A.M. Shumatbaeva, The pH-responsive calix[4]resorcinarene-mPEG conjugates bearing acylhydrazone bonds: Synthesis and study of the potential as supramolecular drug delivery systems, Colloids Surf. A Physicochem. Eng. Asp. 589 (2020) 124453. https://doi.org/10.1016/j.colsurfa.2020.124453

[44] P.F. Gou, W.P. Zhu, Z.Q. Shen, Calixarene-centered amphiphilic A2B2 miktoarm star copolymers based on poly (ε-caprolactone) and poly (ethylene glycol): synthesis and self-assembly behaviors in water, J. Polym. Sci. Part A: Polym. Chem. 48 (2010) 5643-5651. https://doi.org/10.1002/pola.24316

[45] T. Kirila, Thermosensitive star-shaped poly-2-ethyl-2-oxazine. Synthesis, structure characterization, conformation, and self-organization in aqueous solutions, Eur. Polym. J. 120 (2019) 109215. https://doi.org/10.1016/j.eurpolymj.2019.109215

[46] R. Zadmard, Recent progress to construct calixarene-based polymers using covalent bonds: Synthesis and applications, RSC Adv. 10 (2020) 32690-32722. https://doi.org/10.1039/D0RA05707J

[47] S. Suharso, Synthesis of a Novel Calix[4]resorcinarene-Chitosan Hybrid, Oriental Journal of Chemistry, 2018.

[48] Y. Agrawal, J. Pancholi, J. Vyas, Design and synthesis of calixarene, 2009. https://doi.org/10.1002/chin.201028232

[49] U. Trivedi, S. Menon, Y. Agrawal, Polymer supported calix[6]arene hydroxamic acid, a novel chelating resin, Reactive Funct. Polym. 50 (2002) 205-216. https://doi.org/10.1016/S1381-5148(01)00106-7

[50] M. Gidwani, S. Menon, Y. Agrawal, Chelating polycalixarene for the chromatographic separation of Ga (III), In (III) and Tl (III), Reactive Funct. Polym. 53 (2003) 143-156. https://doi.org/10.1016/S1381-5148(02)00169-4

[51] K. Su, Azo-bridged calix[4]resorcinarene-based porous organic frameworks with highly efficient enrichment of volatile iodine, ACS Sustain. Chem. Eng. 6 (2018) 17402-17409. https://doi.org/10.1021/acssuschemeng.8b05203

[52] T. Tilki, An approach to the synthesis of chemically modified bisazocalix[4]arenes and their extraction properties, Tetrahedron 61 (2005) 9624-9629. https://doi.org/10.1016/j.tet.2005.07.078

[53] B. Turner, M. Botoshansky, Y. Eichen, Extended calixpyrroles: Meso-substituted calix [6] pyrroles, Angew. Chem. Int. Ed.. 37 (1998) 2475-2478. https://doi.org/10.1002/(SICI)1521-3773(19981002)37:18<2475::AID-ANIE2475>3.0.CO;2-7

[54] D. Xie, C.D. Gutsche, Synthesis and reactivity of calix[4]arene-based copper complexes, J. Org. Chem. 63 (1998) 9270-9278. https://doi.org/10.1021/jo9810381

[55] D.W. Yoon, H. Hwang, C.H. Lee, Synthesis of a strapped calix[4]pyrrole: Structure and anion binding properties, Ang. Chem. Int. Ed. 41 (2021) 1757-1759. https://doi.org/10.1002/1521-3773(20020517)41:10<1757::AID-ANIE1757>3.0.CO;2-0

[56] D. Siswanta, Calix[4]resorcinarene-chitosan hybrid via amide bond formation, Asian J. Chem. 27 (2015). https://doi.org/10.14233/ajchem.2015.18735

Thermoelectric Polymers: Properties and Applications　　　　Materials Research Forum LLC
Materials Research Foundations 162 (2024) 99-117　　　https://doi.org/10.21741/9781644903018-6

Chapter 6

Thermoelectric Conversion Efficiency and Figure of Merit

Bhavya Padha, Sonali Verma, Sandeep Arya*

Department of Physics, University of Jammu, Jammu, J&K-180006, India

* snp09arya@gmail.com

Abstract

Thermoelectric (TE) materials are useful in renewable energy applications because they can transform waste heat into electricity. To accomplish large-scale thermoelectric applications, materials must have high thermoelectric conversion efficiency, be inexpensive, and operate within a specific temperature range. The high and low temperatures at the junction, as well as a parameter called the figure of merit, are related to the coefficient of performance (COP) of a thermoelectric heat pump and efficiency of a thermoelectric generator. By enhancing electrical behavior and lowering thermal conductivity, a high dimensionless figure of merit (z) value can be achieved.

Keywords

Thermoelectric Materials, Thermoelectric Generator, Thermoelectric Conversion Efficiency, Figure of Merit, Seebeck Coefficient

Contents

1. Introduction

Thermoelectric (TE) materials have ability of converting waste heat to power, and thus are important for renewable energy applications. The task for TE materials in wide range of applications is to find materials having high thermal conversion efficiency, inexpensive, and stable to operate across a wide temperature range. The performance of TE materials are described through the figure of merit,

$$zT = \frac{S^2 \sigma T}{\kappa} \tag{1}$$

Wherein, S, T, κ, and σ are the Seebeck coefficient, absolute temperature, thermal and electrical conductivites, respectively. A high dimensionless figure of merit (z) value is achieved by improving the electrical behaviour and reducing thermal conductivity. Nevertheless, the inverse connection between σ and S makes it difficult to enhance the power factor ($PF = \sigma S^2$). The majority of the notable improvements in z have focused on boosting phonon scattering in material features like dislocations, grain boundaries, interfaces, and precipitates to reduce thermal conductivity [1].

The formula for maximum efficiency (max) of converting heat to electricity by a TE device has been widely employed for determining the suitability of TE materials for devices. Unfortunately, neglecting the temperature dependence of the material characteristics affecting z, the max values differ considerably based on how the average z values are utilized, creating uncertainty in the cases where there is large temperature difference between the warm and cool ends. Engineered power factor $(PF)_{eng}$ and figure of merit $(zT)_{eng}$ depending on the temperature gradient among the warm and cool sides help in predicting accurate conversion efficiency and power output, respectively, managing the exaggerated efficiency employing mean zT values.

A TE generator generates electricity by converting a temperature differential into electricity through TE material. Altenkirch [2] calculated the highest efficiency of a TE generator through constant property model, and its optimized formula has been widely utilized since Loffe [3] presented the ideal situation for maximum efficiency [4].

$$\eta_{\max} = \frac{\Delta T}{T_h} \times \frac{\sqrt{1 + Z \cdot T_{avg}} - 1}{\sqrt{1 + Z \cdot T_{avg}} + \frac{T_c}{T_h}} \tag{2}$$

here T_h and T_c correspond to the temperatures of warm and cool ends, respectively, and ΔT and T_{avg} represent their differential and average values, respectively. **Eq. (2)** represents the TE conversion efficiency as a combination of the Carnot efficiency $(\Delta T/T_H)$ and a reduction parameter corresponding to the figure of merit,

$$z = S^2 \rho^{-1} k^{-1} \tag{3}$$

where S is the Seebeck coefficient, ρ is the electrical resistivity, and k is the thermal conductivity. The dimensionless figure of merit (zT) is described by its peak and average values for better conversion efficiency [5, 6]. The peak efficiency in Eq. (2) is not sufficient when figure of merit depends on temperature. Eq. (2) is valid when figure of merit is independent of temperature and there is a small temperature gradient among the cool and warm ends. In situations where S, ρ, and k depend on temperature, zT depends non-linearly on temperature [7] and temperature difference is also large. Hence, this equation gives inaccurate values. Complex numerical simulations arising from the finite difference approach were used to compute the efficiency whilst considering temperature dependence among the warm and cool ends for overcoming the insufficiency [8].

The efficiency of Eq. (2) is traditionally calculated in one of two different ways:

(i) an integration in terms of temperature,

$$z_{int} = \frac{1}{\Delta T} \int_{T_c}^{T_h} z(T)\, dT \tag{4}$$

(ii) a z value depending on mean temperature,

$$z_{T_{avg}} = z(T_{avg}) \tag{5}$$

Because z is substantially temperature sensitive in certain materials, standard approaches employing average zT typically do not anticipate a realistic efficiency in real working circumstances across a considerable temperature differential. As a result, the traditional efficiency calculation **(Eq. (2))** frequently leads to unrealistically high efficiency predictions. As a result, it is necessary to develop a novel model for predicting energy conversion efficiency for devices working under high temperature differences based on temperature dependent individual TE characteristics [9].

2.   Seebeck coefficient and Thermoelectric figure of merit

The thermoelectric effects that include Seebeck, Peltier as well as Thomson effects and the Joule effect are the two main processes that take place in a TE device. The *Peltier effect* takes place when electric energy is transformed into thermal one, and it exhibits applications in cooling as well as in heating. The thermoelectric cooler (TEC) is the device utilised in such areas [10–12]. In such situation, TE modules are proficient for controlling temperature [13].

The *Seebeck effect* occurs when thermal energy is turned into electric energy, and it shows applications in power production. The thermoelectric generator (TEG) is the device employed in these applications [14, 15]. If a thermal gradient across a conductor produces a voltage at its end points, it is termed as the Seebeck effect. The circuit junctions are composed of two different conductors A and B that are combined together (Fig. 1). The conductors are electrically combined in series whereas thermally they are parallel to each other. T_h represents hot temperature at one junction whereas T_c is the cold one at another junction, with T_h being greater in comparison to T_c. Thermal diffusion causes charge carriers to move over (or against) thermal differences in conductors, resulting in Seebeck effect.

Conductor B

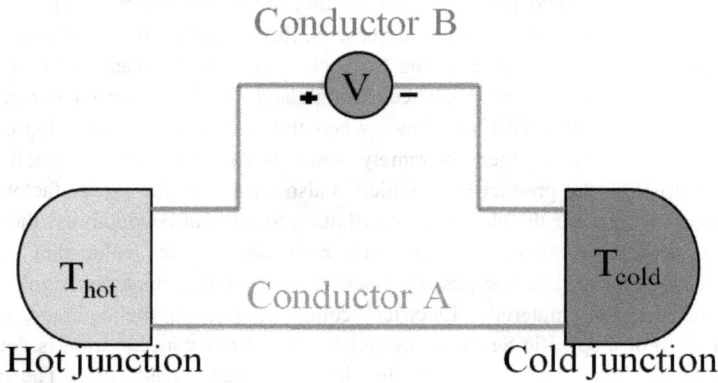

Fig 1: Seebeck effect in case of open circuit.

The Seebeck voltage near circuit junction could be represented as:

$$V = (\alpha_A - \alpha_B) \cdot (T_h - T_c) \tag{6}$$

here α_A is Seebeck coefficient for conductor A and α_B is for conductor B and B (VK^{-1}), $(\alpha_A - \alpha_B) = \alpha_{AB}$ and $(T_h - T_c) = \Delta T$.

2.1   Seebeck coefficient

Efficient thermoelectric materials are semiconductors [16] with doping with acceptor or donor impurity to adjust carrier concentration. The sign of the Seebeck coefficient is likewise determined by these impurities. The quantity that connects the initial thermal differential and the final voltage differential is the Seebeck coefficient of a TE materials or thermopower α_{AB}. A thermoelectric material's Seebeck coefficient is affected by temperature and two other physical transportation factors (thermal and electric conductivity). It estimates the performance of TE materials. Its value varies in the range of μVK^{-1} to mVK^{-1} as well as is affected by the temperature at junctions [17]. The sign is also affected by the type of semiconductor. Moreover, the nature of carrier (e⁻ or h⁺) that conduct electricity affects the sign of the Seebeck coefficient. When e⁻ conducts an electric current, the Seebeck coefficient has a negative sign whereas it is positive if the current carriers are holes h⁺ [18].

Thermoelectric Polymers: Properties and Applications   Materials Research Forum LLC
Materials Research Foundations 162 (2024) 99-117   https://doi.org/10.21741/9781644903018-6

In general, interest is paid towards materials that are dominated by e^- or h^+. Through bipolar conduction, minority carriers decrease Seebeck coefficient as well as enhances thermal conductivity. The value of the Seebeck coefficient increases with decreasing carrier concentration only in the presence of single sign. Seebeck coefficient can grow to the magnitude of ± 1000 µV/K or higher when the energy gap is very high, but the electrical conductivity is then extremely low. When the Seebeck coefficient is around ± 200 µV/K, the product $\alpha 2\ \sigma$ which is also termed as the power factor is at its peak. When we consider the electrical contribution to thermal conductivity, the Seebeck coefficient generating the best power factor approaches to the value that yields the highest figure of merit. The best possible Seebeck coefficient cannot vary significantly as a result of changing materials. Electrical conductivities, on the contrary, can vary substantially. For a specific Seebeck coefficient, the carrier concentration is determined by the effective mass, or more precisely, density-of-states mass (m^*). The electrical conductivity is also affected by the carriers' mobility (μ), which is maximum when the inertial mass is minimum. A high value for the quantity $\mu\ (m^*/m)^{3/2}$ is required. A multi-valley semiconductor exhibits such value. As a result, the most common thermoelectric materials, p-type or n-type Bi_2Te_3 exhibit six valleys. An N-valley semiconductor is the same as having N conductors in the same crystal lattice.

The *Thomson effect* states that heat is released or absorbed inside a conductive substance wherein an electric current passes when there is a thermal gradient across the end points. Thomson heat is represented as:

$$Q = \rho . J^2 - \mu_{AB} \cdot J \cdot \nabla T \tag{7}$$

here $\rho = \dfrac{1}{\sigma}$ represents electric resistivity (Ω m) , σ represents electric conductivity (S m^{-1}), J stands for current density (A m^{-2}), μ_{AB} is Thomson coefficient (VK^{-1}) , whereas ∇T is $\nabla T = \dfrac{dT}{dx}$ is the thermal difference in K. Here $\rho . J^2$ represents joule heating and $\mu_{AB} \cdot J \cdot \nabla T$ represents Thomson heating.

Electric current flowing through a conductor generates heat, which is known as *Joule heating*. In Eq. (7), Joule heating will not vary its sign, whereas Thomson heating does, following J. As a result, the sign conventions for Thomson coefficient are:

- Positive in case of electric charge flowing from the cold to the hot one of a conductor and absorption of heat take place through it;

- Negative in case of the electric charge flowing in reverse direction and there is rejection of the heat from it;
- Null in the situation of current flowing from the hot to the cold side and then in reverse direction as well as the heat does not get absorbed or generated.

The Peltier and Seebeck coefficient, as well as Seebeck and Thomson coefficient, have some relations and such relations are defined as Thomson relations:

$$\pi_{AB} = \alpha_{AB} \cdot T \tag{8}$$

$$\mu_{AB} = T \cdot \frac{d\alpha_{AB}}{dT} \tag{9}$$

2.2 Figure of merit

The temperature at hot and cold junctions as well as a parameter known as the figure of merit (zT) influences the efficiency of TEG and COP of a TE heat pump. The improved value of zT was reported in the middle of the twentieth century. This was achieved by using semiconducting materials with superior electronic characteristics as well as low lattice thermal conduction. Enhancing the scattering of phonons allowed further developments. The quantity power factor, representing a component of the zT which includes the Seebeck coefficient as well as electric conductivity, can also be improved. However, because these values depend on Fermi energy, more progress seems extremely difficult. It is assumed that this will result in limiting zT value.

Practically, the utilization of a zT for single material described as $\alpha^2 \sigma T / \lambda$ is very common. The value of zT of a thermocouple is usually in association with the figures of merits of the two parameters; moreover it will be near to the average of the two parameters. But this condition not always happens. If one branch is a superconductor for example it has zero figure of merit, thus the average will be half that of the other one, while the couple's figure of merit is equivalent to that of the active branch. So, before calculating the couple's figure of merit, we required the ratio of electric to thermal conductivity for every branch. The thermoelectric materials figure of merit zT, represented by **Eq. (1)**, determines the highest efficiency for the energy conversion (either producing power or cooling) at a particular point within the investigated sample [19].

Generally, the figure of merit $zT(T)$ of a material relies on temperature and is obtained from $S(T)$, $\rho(T)$, and $k(T)$ material properties which also depend on temperature.

A proficient thermoelectric generator, on the other hand, must work across a fixed temperature gradient $(T_h - T_c) = \Delta T$, such that the material characteristics will vary from the high temperature side to the low temperature side. Usually, the thermoelectric device ZT is employed to calculate the highest efficiency η of a TE generation system as shown in Eq. 10:

$$\eta = \frac{\Delta T}{T_h} \cdot \frac{\sqrt{1+ZT}-1}{\sqrt{1+ZT}+\frac{T_c}{T_h}} \tag{10}$$

The Carnot factor, $\dfrac{\Delta T}{T_h}$, and the reduced efficiency (η_r) depending on ZT, T_h, as well as T_c limit the generator's overall maximum efficiency. Eq. 10 is generally obtained by assuming that the properties of TE materials (S, ρ, k) are invariable with temperature, that the n and p-type legs are perfectly matched, and the one-dimensional heat flow suffers no additional losses [20].

The material's figure of merit zT for $\left(\dfrac{T_h + T_c}{2}\right)$ and device's figure of merit ZT (assessed in-between T_h and T_c) are identical only in the situation of constant S, ρ, as well as κ. As a result of this observation and relationship of ZT with zT, device ZT is appropriate for describing that the maximum η_r is similar to zT at the average device temperature, and Eq. 10 will be utilized to support the definition of ZT. Because they are readily and frequently misunderstood, a capital Z is used in the device's figure of merit (**Eq. 10**) and a small z is written in the material's figure of merit (**Eq. 1**) [21].

For numerous reasons, such as considerable thermal change of zT and low thermoelectric self compatibility over the temperature window of interest and between the legs, the device ZT can deviate significantly from the material's zT. Many attempts have been made to define ZT as an average of the materials characteristics of the n-leg and p-leg which depends on temperature, with the averages being similar for modest differences in thermoelectric properties between T_h and T_c but significantly different for incompatible thermoelectric segments [22]. Because they assumed various precise temperature dependencies of the TE characteristics and neglect the impacts of TE compatibility, all proposed averaging approaches give inaccurate results [23].

However, calculating device ZT from the properties of thermoelectric material is straightforward and simple. Remember that the goal of a figure of merit is to be helpful, globally accepted, and simple to apply. G. Jeffrey Snyder et al [24] utilized Eq. 10 with some conditions to make it well-defined. Initially, they consider ZT to be a single quantity rather than a product of Z and T. They utilize **Eq. 10** to explain the highest efficiency of a single TE leg (single n- or p-type leg) as the ZT is usually sought for a single material (opposite to a combination n- and p-type pair). Finally, they assume one-dimensional transportation and neglect non-ideal heat and electrical losses. Given a fixed temperature differential $T = T_h - T_c$, the thermoelectric device figure of merit of any material is then determined using the highest efficiency of a single TE leg and **Eq. 10**:

$$zT = \left(\frac{T_h - T_c(1-\eta)}{T_h(1-\eta) - T_c} \right)^2 - 1 \tag{11}$$

Here the highest efficiency η is obtained from the parameters $S(T)$, $\rho(T)$, and $k(T)$ (that depends on thermal difference) between the two sides.

2.3   The dimensionless thermoelectric figure of merit (ZT)

This parameter is utilized to characterize the performance of TE material and the efficiencies of different thermoelectric generators operating at the similar temperatures [25]. The thermal conductivity k, the electric conductivity $\sigma=1/\rho$, as well as Seebeck coefficient α are all physical transportation parameters that influence ZT as shown in Eq.12:

$$zT = \frac{\alpha^2 \cdot T}{\rho \cdot k} = \frac{\alpha^2 \cdot \sigma \cdot T}{k} \tag{12}$$

The term $\alpha^2.\sigma$ is known as the power factor. It is employed to analyze the performance of TE materials. The greater the value of ZT, more efficient is the TE material, and the superior is the thermoelectric generator. The highest ZT in practical applications is around 2, which results in maximum conversion efficiency of around 20% [26].

3.   Thermoelectric conversion efficiency

The electrical efficiency also known as the thermoelectric conversion efficiency of a thermoelectric generator, is the fraction of the electrical output power P transferred to the load with the rate of heat input Q_h absorbed at the high temperature side of the thermoelectric generator. Thus, a thermoelectric generator transforms the rate of heat input Q_h into electrical output power P with thermoelectric conversion efficiency η_{TEG} [27].

$$\eta_{TEG} = \frac{P}{Q_h} \tag{13}$$

Eq. (13) is also represented as

$$\eta_{TEG} = \frac{\eta.R_L.\Delta T.\alpha_{PN}^2}{k.(\eta.R + R_L)^2 + \eta.(R_L.T_h + \eta.R.T).\alpha_{PN}^2} \tag{14}$$

Here T stands for absolute temperature that represents the mean temperature between the low temperature and high temperature sides of the thermoelectric generator represented as $\left(\dfrac{T_h + T_c}{2}\right)$.

The efficiency that corresponds to P_{max} is:

$$\eta_{TEG} = \frac{\Delta T}{4 \cdot z^{-1} + T_h + T} \tag{15}$$

Here $z = \dfrac{\alpha_{PN}^2}{k.R}$ stands for the figure of merit of the thermocouple. Electrical efficiency is maximized in relation to R_L [28] when

$$m = \frac{R_L}{R} = \sqrt{1 + zT} \tag{16}$$

The thermoelectric generator device works as all thermal engines having efficiency smaller in comparison to that of ideal Carnot cycle $\eta_c = \dfrac{T_h - T_c}{T_h} = \dfrac{\Delta T}{T_h} < 1$ [29]:

$$\eta_{TEG_{max}} < \eta_c \tag{17}$$

Here, the Carnot efficiency limits the electrical efficiency, which is represented via incorporating the reduced efficiency (η_r), as:

$$\eta_{TEG_{max}} = \frac{\Delta T}{T_h} \cdot \frac{\sqrt{1+ZT}-1}{\sqrt{1+ZT}+\frac{T_c}{T_h}} = \frac{\Delta T}{T_h} \cdot \frac{m-1}{m+\frac{T_c}{T_h}} = \Delta T \cdot \frac{\eta_r}{T_h} \tag{18}$$

Further, the associated electrical output power is:

$$P_{\eta_{TEG_{max}}} = \frac{\eta \cdot \sqrt{1+ZT}}{R} \cdot \left(\frac{\alpha_{PN} \cdot \Delta T}{\sqrt{1+ZT}} \right) \tag{19}$$

The $\eta_{TEG_{max}}$ = 1% is achieved in case of T_c = 300 K and temperature difference of the order of 20 K [30]. Thermoelectric generator efficiency is greatly influenced by thermoelectric generator operating temperatures (ΔT across the junction), the dimensionless TE figure-of-merit ZT, and thermoelectric generator parameters (area of cross-section, shape and length), as shown in Eq. (18) [31].

The thermoelectric generator efficiency η_{TEG} increases with increasing ΔT (linear relation), and the ratio η_r/T_h is invariable. A thermoelectric generator can operate at around 20% of the Carnot efficiency over a range of high temperatures. The thermoelectric generator efficiency is around 5% and its electrical output power is used at any temperature difference. On introducing the materials with ZT = 10, it results in thermoelectric generators with η_{TEG} = 25% at temperature difference of 300 K [32].

The heat exchanger design has a less impact on thermoelectric waste heat recovery than η TEG. The amount of waste heat transmitted via the thermoelectric couplings is represented by the ratio between thermal efficiency η_t and η TEG [33]:

$$\varepsilon = \frac{\eta_t}{\eta_{TEG}} \tag{20}$$

The highest efficiency, $\eta_{TEG_{max}}$ is related to the thermal gradient ΔT_{TEG} at which the thermoelectric generator operates. The maximum conversion efficiency take place when:

$$\frac{R_L}{R} = \sqrt{1 + z\frac{T_c + T_h}{2}} \qquad (21)$$

4.   Challenges and their possible solutions

The TE materials are selected using a dimensionless quantity β [34] given in Eq. (22)

$$\beta = \left(\frac{k}{e}\right)^2 \frac{\sigma_0 T}{k_l} \qquad (22)$$

where T is absolute temperature, k is Boltzmann's constant, e is electron charge, σ_o is electrical conductivity, and k_l is thermal conductivity. σ_o is related to mobility and effective mass of electron as:

$$\sigma_0 = 2e\mu \left(\frac{2\pi m^* kT}{h^2}\right)^{\frac{3}{2}} \qquad (23)$$

wherein μ is the mobility, m^* is the effective mass of the electron, and h is the Planck's contant. For bismuth telluride, β and zT have values 0.4 and 1, respectively. If $\mu(m^*/m)^{3/2}$ has a value of pure silicon combined to glass-like lattice thermal conductivity, then β would be equal to 3.2, zT equal to 4, and a Seebeck coefficient of about $\pm400\ \mu V\ K^{-1}$.

Upon increasing the values of β further, obtaining high values of zT becomes difficult. This is because the Seebeck coefficient has a linear relationship with Fermi energy while the carrier concentration varies exponentially. The Seebeck coefficient increases with β, but the carrier concentration decreases. Fig. 2 shows variation of zT with β. As zT crosses 1, further increment necessitates much higher values of β. At zT equal to 1, β is equal to 0.4 for bismuth telluride and it rises to 3.2 for zT equal to 4. Doubling β further, makes zT equal to 5.

Some materials have zT close to 1 at room temperature. Therefore, sientists can now create single-stage thermoelectric refrigerators for cooling to around 90 degrees below room temperature. The coefficient of performance of a thermoelectric heat pump working at 30 degree temperature difference among the source and drain is around 1. For the same warm/cool junction temperatures, $zT = 4$ would increase the coefficient of performance to around 3. It could make thermoelectric heat pumps more appealing and a viable alternative to traditional heat pumps presently. The experiment on the French trains [35] was a persuasive example of the benefits of thermoelectric air conditioning. For a train running in Paris, a 20 kW unit was utilized. Over the course of 10 years, the device performed flawlessly.

Electricity generated from a less efficient heat source is perhaps the most important example. The efficiency of a machine with $zT\sim4$ is around 40 percent of an ideal device. This was accomplished through a device having no mobile parts and the ability to operate across a large temperature range [36].

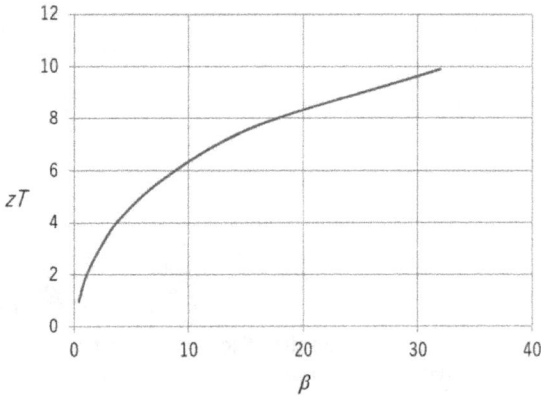

Fig. 2: Variation of zT with β (Reproduced from [36] under Creative Commons Attribution 4.0 License. Copyright 2021, Published by National Institute for Materials Science in partnership with Taylor & Francis Group.).

4.1   Engineering Dimensionless Figure of Merit $(zT)_{eng}$

One of the ways to improve the efficiency and figure of merit of TE devices is to design figure of merit $(zT)_{eng}$.

Thermoelectric efficiency (η) in terms of Seebeck coefficient $[S(T)]$ is given as

$$\eta = \frac{\frac{m}{m+1}}{\frac{m+1}{z_{eng}\Delta T} + \frac{S(T_h)\Delta T}{\int_{T_c}^{T_h} S(T)dT} - \frac{1}{2(m+1)}} \tag{24}$$

where $m = \dfrac{R_L}{R}$, R_L and R being the external and internal electrical resistances, respectively, and $(zT)_{eng} = z_{eng}\Delta T$.

For $m = 1$, Eq. (24) gives the efficiency of a TE material corresponding to the peak output power (P_{max}) as

$$\eta_{P_{max}} = \eta_c \frac{1}{\dfrac{4\eta_c}{(zT)_{eng}} + 2\hat{\alpha} - \dfrac{1}{2}\eta_c} \tag{25}$$

Here η_c is Carnot efficiency, and

$$\hat{\alpha} = \frac{S(T_h)\Delta T}{\displaystyle\int_{T_c}^{T_h} S(T)dT} \tag{26}$$

From Eqs. (24) and (25), it appears that the efficiency (η) depends on $(zT)_{eng}$, hence the maximum efficiency $\eta_{P_{max}}$ requires maximized value of $(zT)_{eng}$. In the entire temperature window, the ΔT-dependent $(zT)_{eng}$ curves of half-Heusler (HH: $Hf_{0.19}Zr_{0.76}Ti_{0.05}$-$CoSb_{0.8}Sn_{0.2}$) and tin selenide (SnSe) exhibiting similar patterns as the actual efficiency assumptions, wherein $(zT)_{eng}$ means the same zT at a particular temperature (T), regardless of value of the highest peak of zT. Despite the fact that SnSe has a substantially higher peak value of zT than HH, its $(zT)_{eng}$ is lower by a factor of three. This is due to the fact that $(zT)_{eng}$ is T-dependent, while zT doesn't have temperature boundary condition's influence or its cumulative effect. The inaccuracy in the zT value is overcome by introducing the effective figure of merit [37],

$$(zT)_{eff} = \frac{\left(\displaystyle\int_{T_c}^{T_h} S(T)dT\right)^2 T_{avg}}{\Delta T \displaystyle\int_{T_c}^{T_h} \rho(T)k(T)dT} \tag{27}$$

Herein, $(zT)_{eff}$ is quite similar to the efficiency excursion.Nonetheless, at exceptionally small ΔT, $(zT)_{eff}$ has implausibly huge values, and at greater T_H, it becomes equivalent to $(zT)_{eng}$. This means that $(zT)_{eng}$ distinguishes the cumulative performance related to a specific temperature differential, whereas $(zT)_{eff}$ denotes the mean performance. Additionally, $(zT)_{eng}$ has a good match with the efficiency estimate, where the vertical axes reflect normalized values like $(zT)_{eng}$ (solid lines), $(zT)_{eff}$ (dashed lines), and η (open symbols).

4.2 Designing power factor and output power density

The output power density (P_d) at the peak efficiency driven by the engineered power factor $(PF)_{eng}$ and optimized resistance ratio (m_{opt}) is given as

$$P_d = \frac{(PF)_{eng}\,\Delta T}{L} \times \frac{m_{opt}}{(1+m_{opt})^2} \tag{28}$$

The output power density is influenced the material qualities, but its efficiency is solely driven by the material characteristics. The output power densities of HH and SnSe at $T_c = 100\ °C$ were also demonstrated [9]. The power density driven using $(PF)_{eng}$ agrees having the numerical value in the range of 1 and 13 percent of relative difference for HH and SnSe, respectively, but the power density based on traditional PF displays 13 and 33 percent of relative difference for HH and SnSe, respectively.

Given the significant difference in thermal conductivity of the materials, the quantity of power generated by HH is ten times that of SnSe in the identical TE leg dimensions. The power density can become analogues by modifying leg dimensions to complement similar input temperature ranges, and yet lowering or boosting leg length creates thermomechanical structural problems, hence a similar leg dimension is regarded for examining the internal behaviour of materials for generating power [9]. The numerical simulation exhibits a comparable tendency and proportional scale to the power density, showing an inherent efficiency for power production of a TE material at practical ΔTs, having significantly distinct trends from power factor (PF). Therefore, $(zT)_{eng}$ and $(PF)_{eng}$ allow comparison of the material's performance at different temperatures without having to undertake comprehensive efficiency and output power generation computations.

Conclusion

The dimensionless figure of merit formula for maximum heat conversion efficiency by a thermoelectric device has been broadly utilized to evaluate the appropriateness of TE materials for practical applications. Unfortunately, maximum efficiency values differ significantly depending on how the average dimensionless figure of merit values are used, enhancing concerns regarding the applicability of dimensionless figure of merit in case of a wide thermal gradient between the hot and cold plates due to the ignorance of temperature dependences of the material properties that influence the dimensionless figure of merit. This chapter revolves around the thermoelectric phenomenon, the figure of merit associated with the material and device and the thermoelectric conversion efficiency. It also focuses on various challenges in this field along with their possible solutions.

References

[1] T.W. Lan, K.H. Su, C.C. Chang, C.L. Chen, M.N. Ou, D.Z. Wu, P.M. Wu, C.Y. Su, M.K. Wu, Y.Y. Chen, Enhancing the figure of merit in thermoelectric materials by adding silicate aerogel, Mater. Today Phys. 13 (2020) 100215. https://doi.org/10.1016/j.mtphys.2020.100215

[2] E. Altenkirch, Uber den Nutzeffekt der Thermosaulen, Phys. Z. 10 (1909) 560-568.

[3] A.F. Ioffe, L.S. Stil'Bans, E.K. Iordanishvili, T.S. Stavitskaya, A. Gelbtuch, G. Vineyard, Semiconductor thermoelements and thermoelectric cooling, Phys. Today. 12 (1959) 42. https://doi.org/10.1063/1.3060810

[4] S.W. Angrist, Direct Energy Conversion, Allyn and Bacon, Inc., Boston, 1965, p. 431.

[5] H.J. Goldsmid, A.R. Sheard, D.A. Wright, The performance of bismuth telluride thermojunctions, Br. J. Appl. Phys. 9 (1958) 365-370. https://doi.org/10.1088/0508-3443/9/9/306

[6] H.J. Wu, L.D. Zhao, F.S. Zheng, D. Wu, Y.L. Pei, X. Tong, M.G. Kanatzidis, J.Q. He, Broad temperature plateau for thermoelectric figure of merit ZT>2 in phase-separated PbTe0.7S0.3, Nat. Commun. 5 (2014) 4515. https://doi.org/10.1038/ncomms5515

[7] S.D. Bhame, D. Pravarthana, W. Prellier, J.G. Noudem, Enhanced thermoelectric performance in spark plasma textured bulk n-type BiTe2.7Se0.3 and p-type

Bi0.5Sb1.5Te3, Appl. Phys. Lett. 102 (2013) 211901.
https://doi.org/10.1063/1.4807771

[8] Q. Zhang, E.K. Chere, K. McEnaney, M. Yao, F. Cao, Y. Ni, S. Chen, C. Opeil, G. Chen, Z. Ren, Enhancement of thermoelectric performance of n-type PbSe by Cr doping with optimized carrier concentration, Adv. Energy. Mater. 5 (2015) 1401977. https://doi.org/10.1002/aenm.201401977

[9] H.S. Kim, W. Liu, G. Chen, C.W. Chu, Z. Ren, Relationship between thermoelectric figure of merit and energy conversion efficiency, Proc. Natl. Acad. Sci. 112 (2015) 8205-8210. https://doi.org/10.1073/pnas.1510231112

[10] R.E. Simons, M.J. Ellsworth, R.C. Chu, An assessment of module cooling enhancement with thermoelectric coolers, J. Heat Transfer Trans. ASME. 127 (2005) 76-84. https://doi.org/10.1115/1.1852496

[11] T.C Cheng, C.H. Cheng, Z.Z. Huang, G.C. Liao, Development of an energy-saving module via a combination of solar cells and thermoelectric coolers for green building applications, Energy. 36 (2011) 133-140. https://doi.org/10.1016/j.energy.2010.10.061

[12] D. Enescu, Thermoelectric refrigeration principle, in: P. Aranguren (Eds.), Bringing Thermoelectricity into Reality, INTECH publishing, 2018, pp. 221-246. https://doi.org/10.5772/intechopen.75439

[13] L.B. Kong, T. Li, H.H. Hng, F. Boey, T. Zhang, S. Li, Waste Energy Harvesting: Mechanical and Thermal Energies, Verlag Berlin Heidelberg, Germany, Springer, 2014. https://doi.org/10.1007/978-3-642-54634-1

[14] D. Champier, J.P. Bedecarrats, T. Kousksou, M. Rivaletto, F. Strub, P. Pignolet, Study of a TE (thermoelectric) generator incorporated in a multifunction wood stove, Energy. 36 (2011) 1518-1526. https://doi.org/10.1016/j.energy.2011.01.012

[15] D. Champier, J.P. Bedecarrats, M. Rivaletto, F. Strub, Thermoelectric power generation from biomass cook stoves, Energy. 35 (2010) 935-942. https://doi.org/10.1016/j.energy.2009.07.015

[16] H.J. Goldsmid, R.W. Douglas, The use of semiconductors in thermoelectric refrigeration, Br. J. Appl. Phys. 5 (1954) 386. https://doi.org/10.1088/0508-3443/5/11/303

[17] E.M. Barber, Thermoelectric Materials: Advances and Applications, NY, USA: Taylor & Francis Group, Pan Stanford, 2015.

[18] G. Neeli, D.K. Behara, M.K. Kumar, State of the art review on thermoelectric materials, Int. J. Sci. Res. 5 (2016) 1833-1844.

[19] G.J. Snyder, Thermoelectric Power Generation: Efficiency and Compatibility, in: D.M. Rowe (Eds.), Thermoelectrics Handbook: Macro to Nano, CRC Press, 2006.

[20] R.R. Heikes, R.W. Ure, Thermoelectricity: Science and Engineering (Interscience, New York, 1961)

[21] G.J. Snyder, E.S. Toberer, Complex thermoelectric materials, Nat. Mater. 7 (2008) 105-114. https://doi.org/10.1038/nmat2090

[22] G.J. Snyder, Application of the compatibility factor to the design of segmented and cascaded thermoelectric generators, Appl. Phys. Lett. 84 (2004) 2436-2438. https://doi.org/10.1063/1.1689396

[23] E. Müller, K. Zabrocki, C. Goupil, G.J. Snyder, W. Seifert, Functionally graded thermoelectric generator and cooler elements, in: D.M. Rowe (Eds.), Thermoelectrics and its Energy Harvesting, CRC Press, 2012.

[24] G.J. Snyder, A.H. Snyder, Figure of merit ZT of a thermoelectric device defined from materials properties, Energy Environ. Sci. 10 (2017) 2280-2283. https://doi.org/10.1039/C7EE02007D

[25] B. Orr, A. Akbarzadeh, M. Mochizuki, R. Singh, A review of car waste heat recovery systems utilizing thermoelectric generators and heat pipes, Appl. Therm. Eng. 101 (2016) 490-495. https://doi.org/10.1016/j.applthermaleng.2015.10.081

[26] B.S. Yilbas, A.Z. Sahin, Thermoelectric device and optimum external load parameter and slenderness ratio, Energy. 35 (2010) 5380-5384. https://doi.org/10.1016/j.energy.2010.07.019

[27] G.J. Snyder, Thermoelectric energy harvesting, in: S. Priya, D.J. Inman (Eds.), Energy Harvesting Technologies. Boston, MA, USA: Springer; 2009) 325-336. https://doi.org/10.1007/978-0-387-76464-1_11

[28] M. Hodes, Optimal pellet geometries for thermoelectric power generation, IEEE Transactions on Components and Packaging 33 (2010) 307-318. https://doi.org/10.1109/TCAPT.2009.2039934

[29] D.M. Rowe, Handbook of Thermoelectrics. Introduction, Boca Raton, Fl, USA: CRC Press, Taylor & Francis Group (1995) 720. https://doi.org/10.1201/9781420049718.ch1

[30] M.E. Kiziroglou, E.M. Yeatman, Materials and techniques for energy harvesting, In: Kilner JA, Skinner SJ, Irvine SJC, Edwards PP, editors. Woodhead Publishing Series in Energy, Functional Materials for Sustainable Energy Applications. Cambridge, UK: Woodhead Publishing; (2012) 541-572. https://doi.org/10.1533/9780857096371.4.539

[31] H. Ali, A.Z. Sahin, B.S. Yilbas, Thermodynamic analysis of a thermoelectric power generator in relation to geometric configuration device pins, Energy Convers. Manag. 78 (2014) 634-640. https://doi.org/10.1016/j.enconman.2013.11.029

[32] O.H. Ando Junior, A.L.O. Maran, N.C. Henao, A review of the development and applications of thermoelectric microgenerators for energy harvesting, Renew. Sustain. Energy Rev. 91 (2018) 376-393. https://doi.org/10.1016/j.rser.2018.03.052

[33] D.T. Crane, G.S. Jackson, Optimization of cross flow heat exchangers for thermoelectric waste heat recovery, Energy Convers. Manag. 45 (2004) 1565-1582. https://doi.org/10.1016/j.enconman.2003.09.003

[34] R.P. Chasmar, R. Stratton, The thermoelectric figure of merit and its relation to thermoelectric generators, J. Electron. Control. 7 (1959) 52-72. https://doi.org/10.1080/00207215908937186

[35] J.G. Stockholm, CRC Handbook of thermoelectric, ed. D.M. Rowe, CRC Press, Boca Raton, Florida, 1994, p. 657.

[36] H.J. Goldsmid, Improving the thermoelectric figure of merit, Sci. Technol. Adv. Mater., 22 (2021) 280-284. https://doi.org/10.1080/14686996.2021.1903816

[37] A. Muto, D. Kraemer, Q. Hao, Z.F. Ren, G. Chen, Thermoelectric properties and efficiency measurements under large temperature differences, Rev. Sci. Instrum. 80 (2009) 093901. https://doi.org/10.1063/1.3212668

Thermoelectric Polymers: Properties and Applications   Materials Research Forum LLC
Materials Research Foundations 162 (2024) 118-143   https://doi.org/10.21741/9781644903018-7

Chapter 7

Other New Thermoelectric Compounds

Uzma Hira*, Adnan Khadim Bhutta and Asifa Safdar

School of Physical Sciences (SPS), University of the Punjab, New Campus, 54590, Lahore, Pakistan

uzma.sps@pu.edu.pk

Abstract

Thermoelectric (TE) compounds have made contributions to solving the energy crisis problem of the globe by providing various sustainable energy solutions. TE materials can transform waste heat of thermal power plants, automobiles, incinerators and domestic cooking stoves into electricity production. TE gadgets consist of n & p-type semiconductors in which temperature changes on the two different sites cause the flow of charges i.e., electrons and holes which produce voltage difference through Seebeck effect. Conventional metal alloys, half-Heuslar, Skutterudite compounds and metal oxides (MO) are considered important TE compounds owing to their high value of electrical conductivity, thermal stability, tunable electron transport, and phonon properties. But due to their rigidity and expensiveness, these are being replaced by polymeric compounds, which show excellent thermoelectric properties and are less expensive as compared to corresponding inorganic materials. In this chapter, the thermal electric properties of various promising n and p-type polymeric compounds are discussed in detail. Moreover, emerging TE applications in different fields of life are also discussed.

Keywords

Thermoelectric Compounds, p-type and n-type Semiconductors, Seebeck Effect, Half-Heuslar and Skutterudite Compounds

Contents

1.   Introduction

Due to the shortage of energy means and continuous depletion of all fossil fuel reserves such as petroleum and natural gas, energy and climatic problems have become global problems. Moreover, annually incomplete usage of energy from the world's total produced energy from different primary energy resources, causes around 60-61 % of the energy of the world is being lost in the form of waste heat [1]. This huge waste of heat demands special consideration and concern for its complete utilization. In this regard, thermoelectric (TE) renewable technology can transform waste heat into electricity production in a completely green/clean ecological way and it can meet the growing energy demand of the rapidly increasing population of the globe. In recent years TE technology has attracted considerable attention from the scientific community [2]. TE modules have a number of advantages as they have no moving parts and involvement of chemicals reaction and therefore are maintenance-free [2].

A TE device consists of a uni-couple, which is comprised of p & n-type materials that are interconnected with the help of electrodes to each other (See Fig. 1A). A TE Module is a working module and it is created by electrically and thermally connecting dozens or hundreds of uni-couples in parallel and series (Fig. 1B). Changing the temperature on one side of the device leads to a voltage gradient and it results in migration of charge carriers (n) from hotter end of the material to the colder end when exposed to a heat source (i.e., Seebeck effect) [3]. In the TE process, a temperature gradient (ΔT) causes charge movement within the material, which produces a voltage gradient (ΔV). The dimensionless thermoelectric diagram of merit is used for measuring the efficacy of thermos-electric materials by using Eq. (1):

$$zT = \frac{\sigma.s^2 T}{\kappa_{Total}} \tag{1}$$

Where σ is termed as electrical conductivity, S represents Seebeck coefficient, κ_{Total} is TC (total thermal conductivity), and absolute temperature represents T. Moreover, the product of s^2 and σ is referred to as PF (Power Factor) [4]. For practical applications, the zT value should be equal to or greater than one.

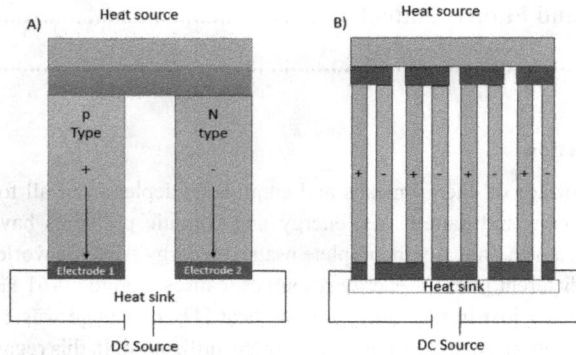

Figure 1.   *A visual representation of Seebeck unicouple (A); thermoelectric module with several unicouples in series and parallel for thermal cooling (B).*

From the last few years, inorganic-based semiconductors-like materials including conventional TE alloys (antimony and bismuth tellurides, etc.) [5,6], skutterudite [7], chalcogenide [8], half-Heusler [9] and inorganic oxide compounds have been investigated as potential TE materials. But, the constituting elements of the above-mentioned compounds have a high cost, toxicity, and less natural abundance [10]. Moreover, manufacturing of these materials requires high synthesis temperatures, which also increases the processing cost of materials. Additionally, inorganic compounds are generally inflexible, consequently, the transformation of waste heat from unevenly molded substances is problematic. All the above-mentioned disadvantages make inorganic materials a complicated choice for capturing waste heat.

1.1   Organic conjugated polymers as promising TE materials

Currently, conjugated polymeric materials are being investigated as an interesting organic TE material since they have the following key characteristics:

- The constituting elements of organic TE polymers have high earth abundance.
- Conjugated polymers can be fabricated at room temperature by using the solution-based approach, which helps in the mass production of TE modules with significantly reduced price.
- It is possible to harvest waste heat from diverse geometries owing to the flexible and light weighted nature of conjugated polymers.
- TE properties can be optimized to a maximum limit easily through structural variation of the compounds.
- The less toxic nature of conjugated polymers makes them desirable as environmentally friendly energy material.

In the course of the last few years, considerable efforts have been carried out for the production of TE compounds, as demonstrated by the increasing number of international publications over the years [11]. The research endeavors on organic-based TE polymeric compounds are in the initial phase. In the beginning, zT values of thermoelectric polymeric materials were mostly low than inorganic materials.

1.2   Power factor (*PF*) optimization

The thermoelectric properties of polymeric compounds are assessed through thermoelectric *PF*. For most of the thermoelectric organic polymers, specifically the class of conjugated polymeric materials, Seebeck coefficient and electrical conductivity have this experimental correlation i.e., (S $\propto \sigma^{-1/4}$) [12]. Therefore, power factor and electrical conductivity also obey the experimental correlation i.e., ($PF \propto \sigma^{1/2}$). Therefore, it is very important for practical application that electrical conductivity values should be high for thermoelectric polymeric materials. Presently, p-type (hole carriers) conjugated polymeric materials are investigated with a good σ value of nearly around 4.0 x 10^3 S/cm [13], with a maximum high σ of around 9.0 x 10^3 S/cm [13]. However, the research on n-type (electron carriers) polymeric materials is significantly slow, because several n-type polymeric compounds possess very high electrical resistivity.

1.3   Design of new potential organic thermoelectric polymers

Detailed knowledge of structure and property correlation is important to design any new high-performing organic thermoelectric polymer. Owing to the disordering nature of conjugated polymeric compounds, their inherent electrical conductivity is naturally inferior. Therefore, the thermoelectric properties of polymeric compounds can be optimized to a little extent. There are different approaches that are used nowadays to fabricate the TE materials of desired properties. Previously tested that significant enabler through decreasing "ohm losses" in the organic semiconducting modules including organic solar cell (OSC) and field effect transistors [14]. Doping is also another good approach, which is extensively used to improve the thermoelectric properties of conjugated polymeric materials [14]. Fig. 2 (upper portion) shows redox doping occurred either via shifting of electron carriers from the doping molecule to the LUMO of polymeric material (reduced doping & n-type), or via HOMO of polymeric material to doping molecule (oxidized doping & p-type). The effectiveness of doping depends on the excellent miscibility of a doping molecule with a polymeric material and reasonable matching of energy states of both for the adequate thermo-dynamic driving force. Moreover, other than the redox-type of doping approach, acid-base-type doping comprises the transmission of cation/hydride ions from dopant molecules to organic TE polymeric compounds (Fig. 2, lower portion), like Lewis acid & base and proton acids-bases. Other emerging advanced doping approaches consist of an ion-exchange substitution method in which a molecular ions-doped afterward addition of preliminary carriers then this dopant is swapped with stable electrolytic/ionic liquids [15].

However, a continuing contest in the plotting of organic TE polymeric compounds and commonly with all TEs is: S and σ are interconnected inversely with n. Consequently, there is tradeoff between S and σ for obtaining high PF. According to the Eq. 2:

$\sigma = n\mu q$ (2)

Where the value of σ depends on both n (carrier concentration) and μ (carriers' mobility). Normally, when doping levels of TE polymeric materials are optimized, a specific quantity of doping element is needed to mix with polymeric material, in order to increase carrier concentration and therefore increase electrical conductivity. However, a high level of n (charge carriers) decreased the Seebeck coefficient. Furthermore, a high quantity of doping may interrupt the shape of films and reduce the carrier's mobility, therefore resulting in a decrease in electrical conductivity. Therefore, S and σ decoupling would permit further increasing the *PF* of TE polymers. In this chapter, few examples of some new potential TE polymeric compounds will be discussed in detail. Additionally, we will also discuss important approaches for decreasing the trade-off linkage between S and σ, and some important applications of thermoelectric polymeric compounds.

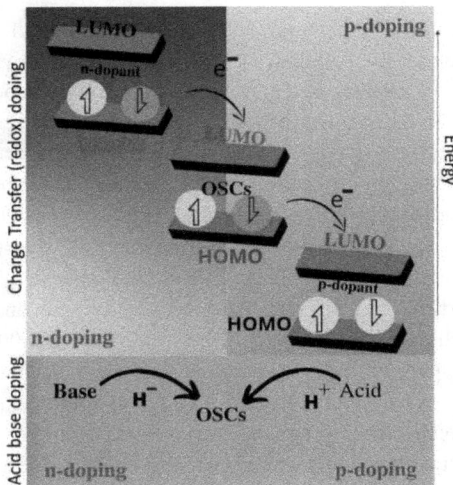

Figure 2.   Representation of redox-type doping, which comprises shifting of electronic carriers to the n-type doping (LUMO)or from the p-type doping(HOMO) of OSC (upper), and acid & base-type doping via shifting of hydride cations (p-type doping) or hydride anions (n-type doping) of OSC (lower).

2.   *p*-type TE polymeric compounds

The progress of *p*-type thermoelectric polymeric compounds can be overviewed since the discovery of poly-acetylene (PA) as a conductive polymer [16]. The doping of iodine in PA causes an increase in the electrical conductivity [17, 18]. The thermoelectric properties of iodine and metal halide-doped PA [19, 20] were examined. However, owing to their reduced stability [54] and negligible solubility in different organic nature solvents, more TE studies about this material were hindered. Afterword, many p-type polymeric materials investigated as interesting TE materials and some of these are listed below:

- PEDOT (Poly(3,4-ethylenedioxythiophene)-based polymeric compounds)

- PBTTT (Poly-thiophene-type semi-crystalline polymeric compounds such as (Poly(3-hexylthiophene) (P3HTP) and poly(3,3′-dialkylquaterthiophene), etc.

- Copolymers like benzothiadiazole(BT) and diketopyrrolopyrrole(DPP)

The TE properties of the above mentioned polymers are explained below in detail.

2.1   Poly (styrenesulfonate): PEDOT as a promising TE material

PEDOT was fabricated in 1980 using monomer EDOT (3, 4-ethylenedioxythiophene). It has limited solubility and is not fusible [21]. An electrical conductivity of around 31 S/cm was obtained for PEDOT hard-pressed pellets. In order to solve processing issues, EDOT monomer was initially oxidized with per-sulfate and then the process of polymerization was carried out with PSS (poly(styrene sulfonate)) solution under aqueous medium, a highly stable PSS and PEDOT aqueous suspension obtained, where poly(styrene sulfonate) helps in dispersion and acts as a dopant in PEDOT [22]. However, PSS increased the processing power of PEDOT, but the insulating nature of PSS in solid-state film limits the electrical properties and molecular packing of PEDOT. Consequently, PSS: PEDOT: The film displays increased electrical resistivity around 0.1−1.0 S/cm and shows thermopower $< 20.0*10^{-6}$V/K, which decreased power factor by the value of $0.001*10^{-6}$W/m·K^{-2} [23]. To solve this problem, the addition of a second doping agent (such as surfactant [24], acid [25], and solvent treatment are required [26]). The action of the second doping agent is on the charge transport mechanism through the improvement of microstructure. For example, Fig. 3(A) displays that ordering and crystal nature along the side planes were improved with adding EG (ethylene glycol) to the aqueous suspension of PSS: PEDOT, which result in a significant increase in carriers mobility from $(4.5 \times 10^{-2}$ to 1.7 cm^2/Vs) [26]. As a result, the electrical conductivity value enlarged up to 8.30×10^2 S/cm in the presence of 3 % ethylene glycol [26]. Fig. 3(B) shows that when PSS: PEDOT films post-treated with H$_2$SO$_4$, a structural reformation is introduced which results in highly arranged polymeric nanofibers with significantly high value of electrical conductivity in the range of 1-4.380 $\times 10^3$ S/cm [27]. Fig. 3(C) shows that PSS: PEDOT suspension was obtained with the addition of 5 % of DMSO (dimethyl sulfoxide) and after that polymeric films were attained through a spin coating process. Next, these polymeric films were again doped with DMSO and hydrazine, and this action is called over coating approach (See Fig. 3(C)). Consequently, the extra PSS was removed and the PEDOT chains so obtained have increased S and decreased σ. However, through carefully optimizing the amount of DMSO N_2H_4 in DMSO, the level of doping can be easily improved and value of Power Factor reached up to $142.0*10^{-6}$ W/m·K2, with σ and thermopower values are 578.0 S/cm and $67.0*10^{-6}$ V/K respectively [28].

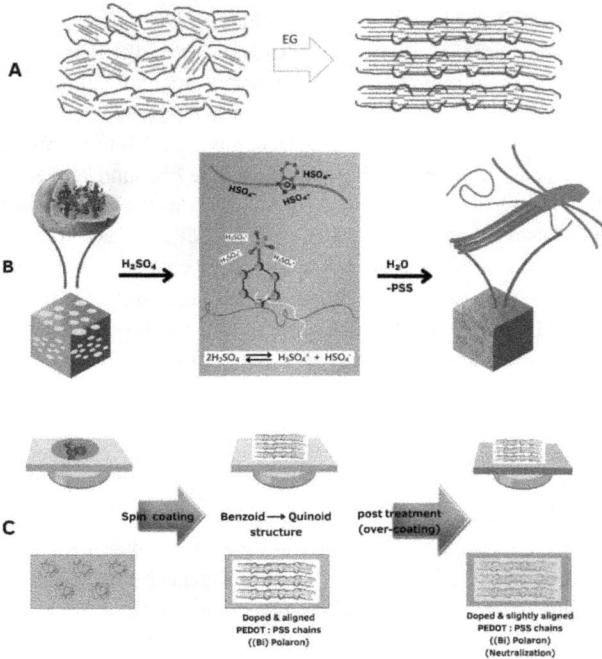

Figure 3. *(A) Presentation of PSS: PEDOT film morphological changes after addition of Ethylene glycol [26] (A). (B) Illustration of an amorphous PSS: PEDOT grain particles (on the left side) are transformed into crystalline PSS: PEDOT nanofibers (on the right side) through a charge-detached transition mechanism (middle) brought with conc. sulfuric acid H2SO4 addition. [27] (B), Representation of the de-doping process of PSS: PEDOT nano-films via over-coating and de-doping approaches (C) [28].*

Moreover, smaller anions were used as dopants instead of PSS in order to decrease the insulating stage, amongst which Tos (tosylate) signifies a very important stage as it permits the improvement of power factor of PEDOT around 3.24×10^2 μW/mK^2 for room temperature TE applications [29] and represented half metallic-type behavior, which is particularly required for TE polymers. Commonly talking, extraordinary development has been made in manufacturing PEDOT class of materials as promising thermoelectric materials from the last few years.

2.1.1 Nano structuring approach in PEDOT family

Nano structuring approach in the above mentioned PEDOT family of compounds can be easily used to enhance TE properties [30]. Chen *et. al.*, [30] fabricated several PEDOT nanostructured materials through chemical method of polymerization in the inverse micro-emulsion. They observed that nanostructuring approach in the PEDOT materials resulted in high carrier's mobility than parent PEDOT compound due to more arranged polymeric layers. PEDOT in nanofiber morphology displayed the σ around 7.14 S/cm, S nearly about $48.0*10^{-6}$V/K and power factor $\sim 16.40*10^{-6}$W/m·K^2. In the form of nanowires, PEDOT was also induced as a filling agent in (Tos or PSS:PEDOT) materials. The nanowire form of PEDOT can alter charge movement via polymeric chains through scattering the lower energy charge carriers. The PEDOT nanowires with 0.2 wt. % loading, the PSS:PEDOT composite materials and nanowires can display a $S \sim 38.9$ µV/K, electrical conductivities values ~ 700.0 S/cm and a power factor of approximately $102.70*10^{-6}$W/m·K^2 [31].

2.1.2 PEDOT/CNT composites

PSS: PEDOT/carbon nanotubes TE composites have been investigated by several groups through different methods [32, 33]. These composites were initially synthesized in 2010 with DMSO, and insulating Arabic gum (AG). The values of electrical conductivities of composite/hybrid materials were affectedly improved due to addition of different functional groups in PSS: PEDOT on the carbon nanotube substrate. Total thermal conductivity value obtained for these composites was very low due to decrease in phonon transfer across tube-tube junction which is initiated by various divergences in the vibrational spectroscopic patterns between PSS: PEDOT and carbon nanotubes. Subsequently, zT value of approximately 0.020 was achieved at room temperature [34, 35].

2.1 Semi-crystalline TE polymeric materials

2.1.1 Polythiophene (PTP) derivatives

Polythiophene (PTP) derivatives had been widely studied due to thermoelectric applications. Conducting polymeric layers can be synthesized through the chemical or electrochemical polymerization process of the corresponding monomeric units. PTP compounds can also be manufactured by adding relevant dopants.

2.1.1.1 Electro-chemical polymerization and TE properties of PTP

The electrochemical polymerization process is a general procedure to synthesize PTP films and their other compounds. The TE properties of these polymeric films, like σ and S, rely on the processing settings. Shinohara *et al.* [36] observed that σ of PTP was more dependent on the synthesis conditions than S. The PTP films exhibited crystalline nature with σ values of ~ 201.0 S/cm and thermoelectric power factor of $\sim 10.30 * 10^{-6}$ W/m·K^2. In one of the research papers, Xu, *et al.* [37] reported the improved electro-chemical polymerizing process for the synthesis of the following compounds as PTP, P3MTP [poly(3-methyl-thiophene)], and P3OTP [poly(3-octyl-thiophene)] through the addition of DTBPD (2,6-di-tert-butylpyridine) into the polymerization mixture. DTBPD functions as a proton catcher, and it can reduce the acidic nature of the reaction mixture and also control side reaction, which is intimated through acidic conditions. Consequently, it enhances the conjugated polymeric length and packing of interchains. They obtained $\sigma \sim 65.9$ S/cm, S around $36.70 * 10^{-6}$ V/K, and power factor approximately $8.880 * 10^{-6}$ W/m·K^2.

2.2.1.2 PTP derivative: P3HTP (Poly (3-hexyl-thiophene-2,5-diyl)

Poly (3-hexyl-thiophene) abbreviated as P3HTP is a polymer from the group of polythiophenes and an organic *p*-semiconducting material. TE Application of these materials depends on appropriate optimization and polymerization processing method, moreover, these are also used in organic photovoltaics, or in field-effect transistors as a semiconducting layer. P3HTP structure is given below [38].

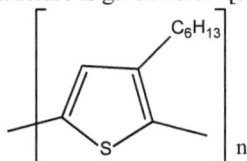

$$C_6H_{13}$$

$$S \qquad n$$

The effective synthesis of P3HTP in 1990 signifies a trend in the improvement of (organic) electronic gadgets [39]. P3HTP is an important semi-crystalline polymeric material, which displays hole carrier mobility around 0.1 cm^2/V*s [40]. According to a well-established pattern from the last twenty years, the carrier transport mechanism of semi-crystalline type polymeric materials rests on various parameters like regio-regularity [40], molecular weight (m.w) [41], and poly-dispersity index [42]. The NOPF6 (nitrosonium hexafluorophosphate) doped P3HTP material shows electrical conductivity around 1.0 S/cm and improved given $PF \sim 0.14 * 10^{-6}$ W/m·K^2 [43]. This was noticed that the presence of PF_6^- ion disturbed the molecular packing of P3HTP, therefore TE properties have been changed. Further, when dipped in the ferric salt solution with Fe(TFSI)$_3$ triflimide anions, the P3HTP mechanically adjustable films were attained, which displayed improved electrical conductivities values around 87 S/cm and a good PF > 20*10^{-6}W/m·K^2 for room temperature TE applications [44]. It is evident from Fig. 4 that F4TCNQ [2,3,5,6-Tetrafluoro-7,7,8,8-tetracyanoquinodimethane] substitution in P3HTP parent compound through ground state carrier transferring process results in the formation of polaron/bi-polaron [45], where the charge carriers transfer from HOMO of P3HTP parent compound to LUMO of F4TCNQ dopant [46]. The F4TCNQ doping in the P3HTP compound through the mixed solution method gives a high conductivity of around 1 to 2 S/cm [47]. The solution mixed doping even with a small amount of doping results in the formation of aggregates, disturbs morphological features and charge carriers mobility, and therefore results in a prominent decrease in conductivity [48]. To solve this issue, the approach of sequential doping was used (Fig. 4). The F4TCNQ spin coating layer on top of P3HTP film generates a significantly very homogenized nano-scale layer, where the doping molecule does not disturb P3HTP crystal structure, and results in n conductivity around 3-9 S/cm [49]. In the next step of the process, F4TCNQ dopant was sublimated so that it can diffuse into the parent P3HTP compound and maintain the crystal structure of the parent compound [50].

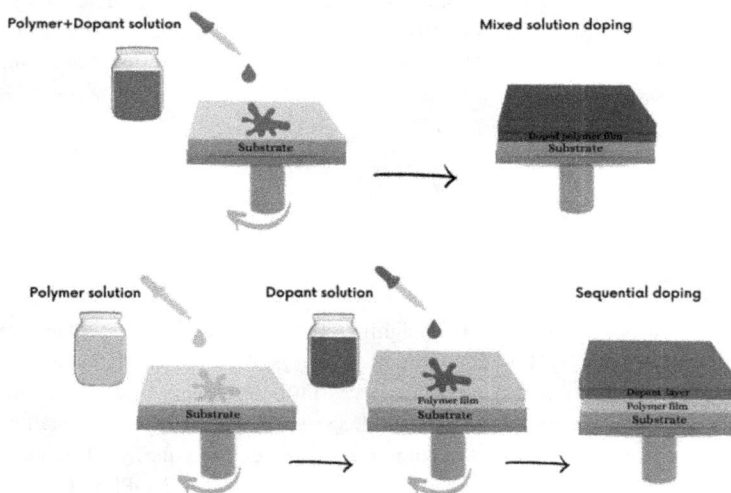

Figure 4.   Representation of solution mixture of dopant and sequential doping.

2.2.1.3 PTP/CNT composites

To create P3HTP/CNT composite material with or without using doping of $FeCl_3$, DCB and chloroform solutions of CNT dispersion and P3HTP solution were directly mixed. $FeCl_3$ doping has a larger effect on the electrical conductivity of these composites. [51, 52]. On the other side, the E (Seebeck coefficient) remained completely unaffected by the concentration of CNT. When CNT loading was between 8–10 wt. percent, there was little effect on the TC of the composite, which results in room temperature $zT >1.02$. P3HTP/multi-walled CNT composites were initially synthesized using a process known as "in-situ polymerization" in 2012. The PTP/multi-walled CNT TE composites were also synthesized using several processes, including solution mixing, in-situ polymerization, and ball milling process [53].

3.   TE *n*-type polymeric compounds

For the advancement of efficient TE gadgets, both types of materials (*n* & *p*-type) are necessary. For this purpose, materials like *n*-type are the most important part of the TE devices as compared to *p*-type. The logic behind this is that p-type materials are less stable in the air. Here, we will discuss some important *n*-type organic polymers.

3.1   Thermoelectric (*n*-Type) Organic polymeric materials

n-type polymer's stability in air or in ambient conditions is associated with LUMO. If the LUMO of (*n*-type) material is at a low level, they exhibit efficient aero stability. For the sake of lowering the LUMO level, an electron-withdrawing group can be introduced. A number of organic polymers are suggested and they can act as *n*-type materials [54-56]. Some of examples are given below:

- Naphthalenetetracarboxylic dianhydride
- Benzotriazole
- Perylene bisimide
- Naphtho thiophene diimide
- Fullerenes
- TBAF(Tetra-n-butylammonium fluoride)

Thermoelectric characteristics of few *n*-type polymeric compounds [57-65] are given in Table 1. Heeger, et al., [60] in 1981 by using an electrochemical method described *n*-type polyethylene and doped this polymer with compound TBAF. But this compound showed less stability. In 2013, Pei and co-workers reported polyphenylene vinylene (PPV) by introducing electron-withdrawing elements like Cl, F etc., [61]. The PF value of this polymer was ~ 28 µw/m·K^2 [62]. The functionality of (n-type) polymers also influenced by the morphology of crystals.

The PPV derivative compounds with an electron-withdrawing group can be used as n-type thermoelectric polymeric compounds. Chabinyc, et al., [66] doped organic polymers with different materials. For example, polymer (P-NDIOD-T$_2$) was doped with compound N-DMBI which exhibited better stability as compared to the PPV derivatives. The PF value for this polymer was ~ 0.6 µw/m·K^2. Conjugated polymers are also used as TE n-type polymers. They have a high solubility rate in organic solvents and their PF value is around 0.43 µw/m·K^2 [67]. Some materials show both n-type as well as p-type behavior, like poly-peri-naphthalene (PPN). They have low bandgap in the range from 0.035 to 0.13 ev.

Table 1. Thermal Electric properties of few n-type polymers are summarized.

Polymer compound	Doping agent	Seebeck coefficient/ (µV/K)	Conductivity (S/cm)	Power Factor (µW/m·K^2)	Thermal conductivity (W/m·K)	zT
PA	Bu$_4$N	−43.5	5	1	-	-
FBDPPV	DMBI	−141	14	28	-	-
FBDPPV	DMBI	−210	6.2	25.5	-	-
P(NDIOD-T2)	DMBI	−850	0.008	0.6	-	-
BBL	TDAE	−101	0.42	0.43	-	-
K$_x$Niett	-	−126	40	66	~0.2	~0.1
K$_x$Niett	-	−90	210	170	0.4-0.5	0.30
CuBHT	-	−4 to −10	750 to 1580	-	-	-

3.2 Transition metals and Organic Hybrid (n-Type) Polymeric materials

Conjugated polymers having metal atom at their back have good air stability and behave like n-type materials [68–72]. A very good example of n-type materials is poly (nickel-ethylenetetrathiolate) (KxNiett). Its properties and electronic structure have been modified by taking counter cations and various central metallic cations. Zhu, et. al., [72] described synthesis of KxNiett films by using electrochemical polymerization. The power factor value ~ 162.0 µW/m·K^2 was estimated, which is greater than all other n-type conjugated polymeric compounds. KxNiett films are also used to make thermoelectric modules. TE modules have a set of 108 KxNiett separate legs display a potential of approximately $577.80*10^{-6}$W/cm^{-2} [73]. Thermoelectric characteristics of polymers having coordinate bonds of Ni-S, e.g. copper benzenehexathiol complex have been reported [74]. The conductivity and S values obtained were around 1500 S/cm and -4 to -10 μV/K, respectively for Cu-BHT.

4.   Recent trends of TE polymeric compounds

Conjugated polymers are not resistant to high temperature and offer poor performance. They are less efficient as compared to inorganic materials in capturing waste heat and providing power. Conjugated polymers provide greater benefits in terms of stretchability, mechanical flexibility, and softness. These benefits are important in harnessing low-grade heat energy into electricity which powers portable/wearable electronic devices. The power consumption of various devices has been shown in Fig. 5 [75].

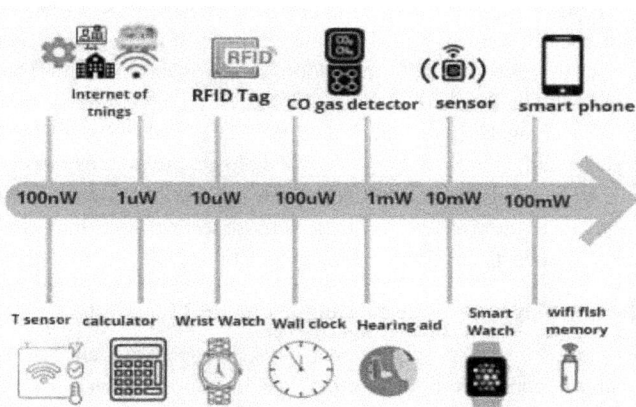

Figure 5.   Electricity utilization related with numerous applications.

For example, an integrated temperature sensor (ITS) showed a consumption of less than 100 nW of power. Internet shows energy consumption at the level of a microwatt. Polymer-based TEGs produce power density in the range of 10-100Wcm^{-2}. By using the radial device model and architecture, we can generate a power density of approximately 1mWcm^{-2}. The use of TE polymers for supplying and harvesting power provides several applications in electronic batteries which overcome the cost problem of batteries [76]. Conjugated polymers will provide significant applications in the future. Here are some examples that explain the potential application of these polymers:

4.1   Self-powered/multi-parameter sensor technology

The first flexible dual-functional sensor capable of simultaneous temperature and pressure detection was announced in 2015. PSS: PEDOT-based sensors which are deposited on a polyurethane frame through the dip-coating method can detect the resistance change in the I-V curve by inducing pressure on one hand, while on the other hand, they will be able to detect the temperature changes by noticing voltage shift in the I-V curve because of the Seebeck effect. Pressure and temperature parameters can also be separated from each other [77].

These sensors are designed as self-powered sensors. Mechanical enhancers like NFC (nano fibrillated cellulose) and PSS: PEDOT-based aerogel such as GOPS (Glycidoxypropyl trimethoxysilane) are also favorable and feasible options for temperature/ pressure double parameter sensors. These sensors experience disturbance in sensing operations due to their inherent transport problems. To overcome the problem, DMSO (dimethyl sulfoxide) vapor method was used to induce all the specific characteristics of PSS: PEDOT aerogels to a distinctive and specific transport-free regime [78]. In the end, dual parameter sensor devices have been achieved. MIECS (mixed ionic electronics conductors) have the ability to sense humidity and temperature gradients as they can transport free electrons and ions. Polymeric MIEC aerogels (nano fibrillated cellulose and PSS: PEDOT) had been fabricated to sense temperature as stable thermo voltage, pressure as resistance shift, and humidity as thermo voltage peak without considering any cross-link. These benefits or achievements provide a new pathway to multipurpose technology which can be harnessed in various safety, diagnostic, and monitoring applications [79].

4.2   Conducting polymeric materials application in TE modules

Stretchable electronics, e.g., stretchable sensors, supercapacitors, and soft batteries have opened a new area of applications, connecting machines and humans in different ways such as implantable and wearable electronic devices [80, 81]. PSS: PEDOT is a crucial material that is important in the area of stretchable electronics [82]. Due to greater E (Young's modulus) and low strain, it becomes impossible to use PSS: PEDOT alone in wearable electronics. The stretchability of PSS: PEDOT can be increased by using many strategies for example volatile surface-acting plasticizers like Zonyl/triton which can be blended with the polymer, and ionic liquids can also be added to the polymeric solution. When we add plasticizers, though it enhances the stretchability, conductivity reduces by one order of magnitude. On the other hand, ionic additives have been proven better which increase both stretchability as well as conductivity [83]. For instance, PSS: PEDOT showed very high conductivity ~ 4100.0 Siemen/centimeters under a hundred percent strain than its unstarched condition, 3100.0 Siemen/Centimeters. It is worth stating that despite all this progress, approaches depend upon their supporting substrates, and after multiple stretching the films are affected out of plane direction by bulk material. Furthermore, the stretchability decreases as the thickness of the film increases. Water-born polyurethane is an effective enhancer discovered recently, conducting elastic components with both water processibility and stretchability were produced as it was blended with PSS: PEDOT [84]. The minimum amount of insulating (water-born) polyurethane is put on in the system to enhance stretchability, which decreases conductivity.

Nowadays thermoelectric module which is intrinsically stretchable in nature has been developed which absorbs heat from the human body. The system has been developed by adding ionic liquids into the water-borne polyurethane and PEDOT which created unique properties i.e., stretchability and conductivity [85]. A composite film can be stretched up to 600 percent and it returns to its original actual form with no hysteresis. Somehow, resistances depending upon strain had a greater effect on the output power when stress is applied to them, but all this opened a new area to work on intrinsically stretchable thermoelectric material. Particularly, these water-based components can be installed on many surfaces which will give environment-friendly and cost-effective products for all fields in daily life.

4.3 Other incipient uses

The photothermal (PT) effect is enchanting more concentration nowadays in harnessing solar energy more efficiently and effectively. Photovoltaic (PV) absorbs the visible light energy that lies within the band gap, while the other part converts the solar energy into thermal energy that could be powered by thermoelectric generators (TEG). TEG/PV integrated devices are considered a new improvement in the research area. The use of inorganic thermal electric materials in integrated TEG/PV devices has resulted in better efficiency [86]. Coupling loss is reduced during the integration of devices. Infrared light can also be used as an illuminator source in integrated devices. Conjugated polymers can also be used as solar absorbents by lowering the coupling loss. For the first time, a solar-based TEG device was demonstrated [87]. For two sun illuminators, a temperature of about 50K was established between the unilluminated sides of PSS: PEDOT and the irradiated side. Such devices prove an output of about 180 nW [88]. Due to this potential, scientists are focusing on the development of more conducting polymers that will harvest solar energy and convert them into electricity. On the other hand, the thermoelectric community is also researching various applications of TE materials in triboelectric TE hybrid nano-generators both mechanical and thermal properties are harvested [89].

Conclusion and Future Outlook

In the past decade, remarkable progress and achievements have been made in the field of thermoelectric polymers. Some potential benefits and applications of thermoelectric polymers have been discussed in this chapter. Thermoelectric polymers are cheaper than inorganic materials. Cost of 25.0g of PSS: PEDOT is less than two hundred dollars. While on the other hand, the cost of bismuth (II) telluride powder is above $300. Thermoelectric polymeric compounds are not efficient enough to harness steam engines. With the advancement of technology, it is predicted that these TE will be used in smart clothes, wearable/ portable electronics, self-powered Internet of Things (IoT), etc. They can also be used in harnessing low-grade heat energy into electricity. Besides this, the organic polymer community requires a number of solutions in this field.

The performance criteria of thermoelectric polymers are not so satisfactory. Seebeck coefficient and electrical conductivity still need some improvement. Operational and air stability are still problems for n-doped TE polymers. Charge-carrying species such as electrons undergo redox reactions with oxygen and water that lose their performance. Poor stability is the biggest hurdle to the advancement of the n-type TE polymers. The aero stability of the n-type TE compounds can be increased by lowering the LUMO level. In recent research, when O_2 of PNDI (Phenylnaphthalenediimide)-based polymer was replaced with S (sulfur), thiolation of the reaction decreased LUMO by a voltage of 0.20 eV which leads to successful air stability and better conductivity. The same polymer shows lower electrical conductivity after 16-17h exposure to ambient conditions.

Similarly, BDOPV, an electron-deficient polymer after substitution with dichloro, secondly copolymerization with a weak CITVT produces a highly stable conductive polymer with conductivity of 4.9 S/cm. For doped polymers, de-doping is a major concern. Self-capsulation is used to harness air stability in this regard. This phenomenon produces stable n-type TE polymers with low LUMO.

Besides all these polymers, conjugated polymers are a hot research topic nowadays due to their mechanical flexibility, thermal properties, and softness. With knowledge of molecular structures and better research areas, the scientific community will be able to discover new advancements from conducted polymers.

References

[1] T. Cao, X.-L. Shi, J. Zou, Z.-G. Chen, Advances in conducting polymer-based thermoelectric materials and devices, Microstructures. 1 (2021) 1-33. https://doi.org/10.20517/microstructures.2021.06

[2] I. Petsagkourakis, E. Pavlopoulou, E. Cloutet, Y.F. Chen, X. Liu, M. Fahlman, M. Berggren, X. Crispin, S. Dilhaire, G. Fleury, G. Hadziioannou, Correlating the Seebeck coefficient of thermoelectric polymer thin films to their charge transport mechanism, Org. Electron. 52 (2018) 335-341. https://doi.org/10.1016/j.orgel.2017.11.018

[3] B.T. McGrail, A. Sehirlioglu, E. Pentzer, Polymer composites for thermoelectric applications, Angew. Chemie - Int. Ed. 54 (2015) 1710-1723. https://doi.org/10.1002/anie.201408431

[4] K. Yusupov, A. Zakhidov, S. You, S. Stumpf, P.M. Martinez, A. Ishteev, A. Vomiero, V. Khovaylo, U. Schubert, Influence of oriented CNT forest on thermoelectric properties of polymer-based materials, J. Alloys Compd. 741 (2018) 392-397. https://doi.org/10.1016/j.jallcom.2018.01.010

[5] B. Poudel, Q. Hao, Y. Ma, Y. Lan, A. Minnich, B. Yu, X. Yan, D. Wang, A. Muto, D. Vashaee, X. Chen, J. Liu, M.S. Dresselhaus, G. Chen, Z. Ren, High-thermoelectric performance of nanostructured bismuth antimony telluride bulk alloys, Science. 320 (2008) 634-638. https://doi.org/10.1126/science.1156446

[6] D. Wright, Thermoelectric properties of bismuth telluride and its alloys, Nature. 181 (1958) 834-834. https://doi.org/10.1038/181834a0

[7] L. Yang, J.S. Wu, L.T. Zhang, Microstructure evolvements of a rare-earth filled skutterudite compound during annealing and spark plasma sintering, Mater. Des. 25 (2004) 97-102. https://doi.org/10.1016/j.matdes.2003.10.005

[8] B. Gahtori, S. Bathula, K. Tyagi, M. Jayasimhadri, A.K. Srivastava, S. Singh, R.C. Budhani, A. Dhar, Giant enhancement in thermoelectric performance of copper selenide by incorporation of different nanoscale dimensional defect features, Nano Energy. 13 (2015) 36-46. https://doi.org/10.1016/j.nanoen.2015.02.008

[9] T. Zhu, C. Fu, H. Xie, Y. Liu, X. Zhao, High efficiency half-heusler thermoelectric materials for energy harvesting, Adv. Energy Mater. 5 (2015) 1-13. https://doi.org/10.1002/aenm.201500588

[10] U. Hira, J.W.G. Bos, A. Missyul, F. Fauth, N. Pryds, F. Sher, $Ba_2-xBi_xCoRuO_6$ ($0.0 \leq x \leq 0.6$) hexagonal double-perovskite-type oxides as promising p-type thermoelectric materials, Inorg. Chem. 60 (2021) 17824-17836. https://doi.org/10.1021/acs.inorgchem.1c02442

[11] D. Beretta, N. Neophytou, J.M. Hodges, M.G. Kanatzidis, D. Narducci, M.M. Gonzalez, M. Beekman, B. Balke, G. Cerretti, W. Tremel, A. Zevalkink, A.I.

Hofmann, C. Müller, B. Dörling, M.C. Quiles, M. Caironi, Thermoelectrics: From history, a window to the future, Mater. Sci. Eng. Reports. 138 (2019) 210-255. https://doi.org/10.1016/j.mser.2018.09.001

[12] E.M. Thomas, B.C. Popere, H. Fang, M.L. Chabinyc, R.A. Segalman, Role of disorder induced by doping on the thermoelectric properties of semiconducting polymers, Chem. Mater. 30 (2018) 2965-2972. https://doi.org/10.1021/acs.chemmater.8b00394

[13] A.C. Hinckley, S.C. Andrews, M.T. Dunham, A. Sood, M.T. Barako, S. Schneider, M.F. Toney, K.E. Goodson, Z. Bao, Achieving high thermoelectric performance and metallic transport in solvent-sheared PEDOT:PSS, Adv. Electron. Mater. 7 (2021) 1-9. https://doi.org/10.1002/aelm.202001190

[14] W. Zhao, J. Ding, Y. Zou, C.A. Di, D. Zhu, Chemical doping of organic semiconductors for thermoelectric applications, Chem. Soc. Rev. 49 (2020) 7210-7228. https://doi.org/10.1039/D0CS00204F

[15] Y. Yamashita, J. Tsurumi, M. Ohno, R. Fujimoto, S. Kumagai, T. Kurosawa, T. Okamoto, J. Takeya, S. Watanabe, Efficient molecular doping of polymeric semiconductors driven by anion exchange, Nature. 572 (2019) 634-638. https://doi.org/10.1038/s41586-019-1504-9

[16] M. Audenaert, G. Gusman, R. Deltour, Electrical conductivity of I2-doped polyacetylene, Phys. Rev. B. 24 (1981) 7380-7382. https://doi.org/10.1103/PhysRevB.24.7380

[17] © 1987 Nature Publishing Group, (1987)

[18] H. Naarmann, N. Theophilou, New process for the production of metal-like, stable polyacetylene, Synth. Met. 22 (1987) 1-8. https://doi.org/10.1016/0379-6779(87)90564-9

[19] H. Kaneko, T. Ishiguro, A. Takahashi, J. Tsukamoto, Magnetoresistance and thermoelectric power studies of metal-nonmetal transition in iodine-doped polyacetylene, Synth. Met. 57 (1993) 4900-4905. https://doi.org/10.1016/0379-6779(93)90836-L

[20] Y.W. Park, C.O. Yoon, B.C. Na, H. Shirakawa, K. Akagi, Metallic properties of transition metal halides doped polyacetylene: The soliton liquid state, Synth. Met. 41 (1991) 27-32. https://doi.org/10.1016/0379-6779(91)90989-I

[21] H. Gerhard, J. Friedrich, Poly(alkylenedioxythiophene)s - new, very stable conducting polymers, Adv. Mater. 4 (1992) 116-118. https://doi.org/10.1002/adma.19920040213

[22] F. Jonas, J.T. Morrison, 3,4-Polyethylenedioxythiophene (PEDT): Conductive coatings technical applications and properties, Synth. Met. 85 (1997) 1397-1398. https://doi.org/10.1016/S0379-6779(97)80290-1

[23] C. Ph, Significant different conductivities of the two grades of poly(3,4-ethylenedioxythiophene):poly(styrenesulfonate), Clevios P and Clevios PH1000, Arising from Different Molecular Weights, (2012). https://doi.org/10.1021/am300881m

[24] C.M. Palumbiny, J. Schlipf, A. Hexemer, C. Wang, P.M. Buschbaum, The morphological power of soap: How surfactants lower the sheet resistance of PEDOT:PSS by strong impact on inner film structure and molecular interface orientation, Adv. Electron. Mater. 2 (2016) 1-9. https://doi.org/10.1002/aelm.201500377

[25] S.R.S. Kumar, N. Kurra, H.N. Alshareef, Enhanced high temperature thermoelectric response of sulphuric acid treated conducting polymer thin films, J. Mater. Chem. C. 4 (2015) 215-221. https://doi.org/10.1039/C5TC03145A

[26] Q. Wei, M. Mukaida, Y. Naitoh, T. Ishida, Morphological change and mobility enhancement in PEDOT:PSS by adding co-solvents, Adv. Mater. 25 (2013) 2831-2836. https://doi.org/10.1002/adma.201205158

[27] N. Kim, S. Kee, S.H. Lee, B.H. Lee, Y.H. Kahng, Y.R. Jo, B.J. Kim, K. Lee, Highly conductive PEDOT:PSS nanofibrils induced by solution-processed crystallization, Adv. Mater. 26 (2014) 2268-2272. https://doi.org/10.1002/adma.201304611

[28] H. Park, S.H. Lee, F.S. Kim, H.H. Choi, I.W. Cheong, J.H. Kim, Enhanced thermoelectric properties of PEDOT:PSS nanofilms by a chemical dedoping process, J. Mater. Chem. A. 2 (2014) 6532-6539. https://doi.org/10.1039/C3TA14960A

[29] O. Bubnova, Z.U. Khan, A. Malti, S. Braun, M. Fahlman, M. Berggren, X. Crispin, Optimization of the thermoelectric figure of merit in the conducting polymer poly(3,4-ethylenedioxythiophene), Nat. Mater. 10 (2011) 429-433. https://doi.org/10.1038/nmat3012

[30] X. Hu, G. Chen, X. Wang, H. Wang, Tuning thermoelectric performance by nanostructure evolution of a conducting polymer, J. Mater. Chem. A. 3 (2015) 20896-20902. https://doi.org/10.1039/C5TA07381B

[31] K. Arlauskas, M. Viliu, K. Genevic, G. Jus, H. Stubb, Charge transport in π-conjugated polymers, Phys. Rev. B. 62 (2000) 235-238.

[32] C. Yu, K. Choi, L. Yin, J.C. Grunlan, Light-weight flexible carbon nanotube based organic composites with large thermoelectric power factors, ACS Nano. 5 (2011) 7885-7892. https://doi.org/10.1021/nn202868a

[33] K. Choi, C. Yu, Highly doped carbon nanotubes with gold nanoparticles and their influence on electrical conductivity and thermopower of nanocomposites, PLOS One. 7 (2012). https://doi.org/10.1371/journal.pone.0044977

[34] D. Kim, Y. Kim, K. Choi, J. Grunlan, C. Yu, Improved thermoelectric behavior of nanotube-filled polymer composites with poly(3,4-ethylenedioxythiophene) poly(styrenesulfonate), ACS Nano. 4 (2010) 513-523. https://doi.org/10.1021/nn9013577

[35] G.P. Moriarty, S. De, P.J. King, U. Khan, M. Via, J.A. King, J.N. Coleman, J.C. Grunlan, Thermoelectric behavior of organic thin film nanocomposites, J. Polym. Sci. Part B Polym. Phys. 51 (2013) 119-123. https://doi.org/10.1002/polb.23186

[36] I. Imae, R. Akazawa, Y. Harima, Seebeck coefficients of regioregular poly(3-hexylthiophene) correlated with doping levels, Phys. Chem. Chem. Phys. 20 (2018) 738-741. https://doi.org/10.1039/C7CP07114K

[37] Y. Hu, H. Shi, H. Song, C. Liu, J. Xu, L. Zhang, Q. Jiang, Effects of a proton scavenger on the thermoelectric performance of free-standing polythiophene and its derivative films, Synth. Met. 181 (2013) 23-26. https://doi.org/10.1016/j.synthmet.2013.08.006

[38] L.A. Kehrer, S. Winter, R. Fischer, C. Melzer, H.V. Seggern, Temporal and thermal properties of optically induced instabilities in P3HT field-effect transistors, Synth. Met. 161 (2012) 2558-2561. https://doi.org/10.1016/j.synthmet.2011.08.007

[39] Z. Bao, A. Dodabalapur, A.J. Lovinger, Soluble and processable regioregular poly(3-hexylthiophene) for thin film field-effect transistor applications with high mobility, Appl. Phys. Lett. 69 (1996) 4108-4110. https://doi.org/10.1063/1.117834

[40] Y. Du, J. Chen, X. Liu, C. Lu, J. Xu, B. Paul, P. Eklund, Flexible n-type tungsten carbide/polylactic acid thermoelectric composites fabricated by additive manufacturing, Coatings. 8 (2018). https://doi.org/10.3390/coatings8010025

[41] H. Sirringhaus, N. Tessler, R.H. Friend, Integrated optoelectronic devices based on conjugated polymers, Science. 280 (1998) 1741-1744. https://doi.org/10.1126/science.280.5370.1741

[42] R.J. Kline, M.D. McGehee, E.N. Kadnikova, J. Liu, J.M.J. Fréchet, M.F. Toney, Dependence of regioregular poly(3-hexylthiophene) film morphology and field-effect mobility on molecular weight, Macromolecules. 38 (2005) 3312-3319. https://doi.org/10.1021/ma047415f

[43] Y. Xuan, X. Liu, S. Desbief, P. Leclère, M. Fahlman, R. Lazzaroni, M. Berggren, J. Cornil, D. Emin, X. Crispin, Thermoelectric properties of conducting polymers: The case of poly(3-hexylthiophene), Phys. Rev. B - Condens. Matter Mater. Phys. 82 (2010) 1-9. https://doi.org/10.1103/PhysRevB.82.115454

[44] Q. Zhang, Y. Sun, W. Xu, D. Zhu, Thermoelectric energy from flexible P3HT films doped with a ferric salt of triflimide anions, Energy Environ. Sci. 5 (2012) 9639-9644. https://doi.org/10.1039/c2ee23006b

[45] H. Méndez, G. Heimel, S. Winkler, J. Frisch, A. Opitz, K. Sauer, B. Wegner, M. Oehzelt, C. Röthel, S. Duhm, D. Többens, N. Koch, I. Salzmann, Charge-transfer crystallites as molecular electrical dopants, Nat. Commun. 6 (2015). https://doi.org/10.1038/ncomms9560

[46] C. Wang, D.T. Duong, K. Vandewal, J. Rivnay, A. Salleo, Optical measurement of doping efficiency in poly(3-hexylthiophene) solutions and thin films, Phys. Rev. B - Condens. Matter Mater. Phys. 91 (2015) 1-7. https://doi.org/10.1103/PhysRevB.91.085205

[47] D.T. Duong, C. Wang, E. Antono, M.F. Toney, A. Salleo, The chemical and structural origin of efficient p-type doping in P3HT, Org. Electron. 14 (2013) 1330-1336. https://doi.org/10.1016/j.orgel.2013.02.028

[48] G. Zuo, Z. Li, O. Andersson, H. Abdalla, E. Wang, M. Kemerink, Molecular doping and trap filling in organic semiconductor host-guest systems, J. Phys. Chem. C. 121 (2017) 7767-7775. https://doi.org/10.1021/acs.jpcc.7b01758

[49] M.T. Fontana, D.A. Stanfield, D.T. Scholes, K.J. Winchell, S.H. Tolbert, B.J. Schwartz, Evaporation vs solution sequential doping of conjugated polymers: F4TCNQ doping of micrometer-thick P3HT films for thermoelectrics, J. Phys. Chem. C. 123 (2019) 22711-22724. https://doi.org/10.1021/acs.jpcc.9b05069

[50] E. Lim, K.A. Peterson, G.M. Su, M.L. Chabinyc, Thermoelectric Properties of poly(3-hexylthiophene) (P3HT) doped with 2,3,5,6-Tetrafluoro-7,7,8,8-tetracyanoquinodimethane (F4TCNQ) by vapor-phase infiltration, Chem. Mater. 30 (2018) 998-1010. https://doi.org/10.1021/acs.chemmater.7b04849

[51] C. Bounioux, P.D. Chao, M.C. Quiles, M.S.M. González, A.R. Goñi, R.Y. Rozen, C. Müller, Thermoelectric composites of poly(3-hexylthiophene) and carbon nanotubes with a large power factor, Energy Environ. Sci. 6 (2013) 918-925. https://doi.org/10.1039/c2ee23406h

[52] W. Li, Z. Zhou, W. Zhou, H. Li, X. Zhao, G. Wang, G. Sun, Q. Xin, Preparation and characterization of Pt/C cathode catalysts for direct methanol fuel cells effect of different preparation and treatment methods, Chinese J. Catal. 24 (2003) 465-470.

[53] L. Wang, X. Jia, D. Wang, G. Zhu, J. Li, Preparation and thermoelectric properties of polythiophene/multiwalled carbon nanotube composites, Synth. Met. 181 (2013) 79-85. https://doi.org/10.1016/j.synthmet.2013.08.011

[54] J.L. Banal, J. Subbiah, H. Graham, J.K. Lee, K.P. Ghiggino, W.W.H. Wong, Electron deficient conjugated polymers based on benzotriazole, Polym. Chem. 4 (2013) 1077-1083. https://doi.org/10.1039/C2PY20850D

[55] R. Schmidt, J.H. Oh, Y. Sen Sun, M. Deppisch, A.M. Krause, K. Radacki, H. Braunschweig, M. Könemann, P. Erk, Z. Bao, F. Würthner, High-performance air-stable n-channel organic thin film transistors based on halogenated perylene bisimide semiconductors, J. Am. Chem. Soc. 131 (2009) 6215-6228. https://doi.org/10.1021/ja901077a

[56] Y. Fukutomi, M. Nakano, J.Y. Hu, I. Osaka, K. Takimiya, Naphthodithiophenediimide (NDTI): Synthesis, structure, and applications, J. Am. Chem. Soc. 135 (2013) 11445-11448. https://doi.org/10.1021/ja404753r

[57] H. Park, S.H. Lee, F.S. Kim, H.H. Choi, I.W. Cheong, J.H. Kim, Enhanced thermoelectric properties of PEDOT:PSS nanofilms by a chemical dedoping process, J. Mater. Chem. A. 2 (2014) 6532-6539. https://doi.org/10.1039/C3TA14960A

[58] M. Nakano, I. Osaka, K. Takimiya, Naphthodithiophene diimide (NDTI)-based semiconducting copolymers: From ambipolar to unipolar N-type polymers, Macromolecules. 48 (2015) 576-584. https://doi.org/10.1021/ma502306f

[59] F. Paquin, J. Rivnay, A. Salleo, N. Stingelin, C. Silva, Multi-phase semicrystalline microstructures drive exciton dissociation in neat plastic semiconductors, J. Mater. Chem. C. 3 (2015) 10715-10722. https://doi.org/10.1039/C5TC02043C

[60] D. Moses, J. Chen, A. Denenstein, M. Kaveh, T.C. Chung, A.J. Heeger, A.G. MacDiarmid, Y.W. Park, Inter-soliton electron hopping transport in trans-(CH)x, Solid State Commun. 40 (1981) 1007-1010. https://doi.org/10.1016/0038-1098(81)90055-7

[61] T. Lei, J.H. Dou, X.Y. Cao, J.Y. Wang, J. Pei, Electron-deficient poly(p-phenylene vinylene) provides electron mobility over 1 cm2 V-1 s-1 under ambient conditions, J. Am. Chem. Soc. 135 (2013) 12168-12171. https://doi.org/10.1021/ja403624a

[62] X. Huang, P. Sheng, Z. Tu, F. Zhang, J. Wang, H. Geng, Y. Zou, C.A. Di, Y. Yi, Y. Sun, W. Xu, D. Zhu, A two-dimensional π-d conjugated coordination polymer with extremely high electrical conductivity and ambipolar transport behaviour, Nat. Commun. 6 (2015) 6-13. https://doi.org/10.1038/ncomms8408

[63] W. Ma, K. Shi, Y. Wu, Z.Y. Lu, H.Y. Liu, J.Y. Wang, J. Pei, Enhanced molecular packing of a conjugated polymer with high organic thermoelectric power factor, ACS Appl. Mater. Interfaces. 8 (2016) 24737-24743. https://doi.org/10.1021/acsami.6b06899

[64] Z. Chen, Y. Zheng, H. Yan, A. Facchetti, Naphthalenedicarboximide- vs perylenedicarboximide-based copolymers, synthesis and semiconducting properties in bottom-gate N-channel organic transistors, J. Am. Chem. Soc. 131 (2009) 8-9. https://doi.org/10.1021/ja805407g

[65] Y. Sun, P. Sheng, C. Di, F. Jiao, W. Xu, D. Qiu, D. Zhu, Organic thermoelectric materials and devices based on p- and n-type poly (metal 1,1,2,2-ethene tetrathionate), Adv. Mater. 24 (2012) 932-937. https://doi.org/10.1002/adma.201104305

[66] Y. Wang, M. Nakano, T. Michinobu, Y. Kiyota, T. Mori, K. Takimiya, Naphthodithiophenediimide-Benzobisthiadiazole-based polymers: Versatile n-type materials for field-effect transistors and thermoelectric devices, Macromolecules. 50 (2017) 857-864. https://doi.org/10.1021/acs.macromol.6b02313

[67] J. Chen, J. Zhang, Y. Zou, W. Xu, D. Zhu, PPN (poly-: Peri-naphthalene) film as a narrow-bandgap organic thermoelectric material, J. Mater. Chem. A. 5 (2017) 9891-9896. https://doi.org/10.1039/C7TA02431B

[68] J.R. Reynolds, C.P. Lillya, J.C.W. Chien, Intrinsically electrically conducting poly(metal tetrathiooxalates), Macromolecules. 20 (1987) 1184-1191. https://doi.org/10.1021/ma00172a003

[69] J.R. Reynolds, C.A. Jolly, S. Krichene, P. Cassoux, C. Faulmann, Poly (metal tetrathiooxalates): A structural and charge-transport study, Synth. Met. 31 (1989) 109-126. https://doi.org/10.1016/0379-6779(89)90631-0

[70] K. Oshima, Y. Shiraishi, N. Toshima, Novel nanodispersed polymer complex, poly(nickel 1,1,2,2-ethene tetrathionate): Preparation and hybridization for n-type of

organic thermoelectric materials, Chem. Lett. 44 (2015) 1185-1187.
https://doi.org/10.1246/cl.150328

[71] Y. Sun, J. Zhang, L. Liu, Y. Qin, Y. Sun, W. Xu, D. Zhu, Optimization of the
thermoelectric properties of poly(nickel-ethylene tetra thiolate) synthesized via
potentiostatic deposition, Sci. China Chem. 59 (2016) 1323-1329.
https://doi.org/10.1007/s11426-016-0175-9

[72] D. de C. C. Faulmann, J. Chahine, K. Jacob, Y. Coppel, L. Valade, Nickel ethylene
tetrathiolate polymers as nanoparticles: a new synthesis for future applications?, J.
Nanopart. Res. 15 (2013) 1586. https://doi.org/10.1007/s11051-013-1586-5

[73] C. Di L. Liu, Y. Sun, W. Li, J. Zhang, X. Huang, Z. Chen, Y. Sun, M.C.F. 2017. W.
Xu, D. Zhu, No Title, (n.d.).

[74] X. Huang, P. Sheng, Z. Tu, F. Zhang, J. Wang, H. Geng, Y. Zou, C.A. Di, Y. Yi, Y.
Sun, W. Xu, D. Zhu, A two-dimensional π-d conjugated coordination polymer with
extremely high electrical conductivity and ambipolar transport behaviour, Nat.
Commun. 6 (2015) 6-13. https://doi.org/10.1038/ncomms8408

[75] A.K. Menon, S.K. Yee, Design of a polymer thermoelectric generator using radial
architecture, J. Appl. Phys. 119 (2016). https://doi.org/10.1063/1.4941101

[76] M. Mukaida, K. Kirihara, Q. Wei, Enhanced power output in polymer thermoelectric
devices through thermal and electrical impedance matching, ACS Appl. Energy Mater.
2 (2019) 6973-6978. https://doi.org/10.1021/acsaem.9b01342

[77] A. Malti, J. Edberg, H. Granberg, Z.U. Khan, J.W. Andreasen, X. Liu, D. Zhao, H.
Zhang, Y. Yao, J.W. Brill, I. Engquist, M. Fahlman, L. Wågberg, X. Crispin, M.
Berggren, An organic mixed ion-electron conductor for power electronics, Adv. Sci. 3
(2015) 1-9. https://doi.org/10.1002/advs.201500305

[78] Z.U. Khan, J. Edberg, M.M. Hamedi, R. Gabrielsson, H. Granberg, L. Wågberg, I.
Engquist, M. Berggren, X. Crispin, Thermoelectric polymers and their elastic aerogels,
Adv. Mater. 28 (2016) 4556-4562. https://doi.org/10.1002/adma.201505364

[79] S. Han, N.U.H. Alvi, L. Granlöf, H. Granberg, M. Berggren, S. Fabiano, X. Crispin,
A multiparameter pressure-temperature-humidity sensor based on mixed ionic-
electronic cellulose aerogels, Adv. Sci. 6 (2019).
https://doi.org/10.1002/advs.201802128

[80] M.S. White, M. Kaltenbrunner, E.D. Głowacki, K. Gutnichenko, G. Kettlgruber, I.
Graz, S. Aazou, C. Ulbricht, D.A.M. Egbe, M.C. Miron, Z. Major, M.C. Scharber, T.
Sekitani, T. Someya, S. Bauer, N.S. Sariciftci, Ultrathin, highly flexible and

stretchable PLEDs, Nat. Photonics. 7 (2013) 811-816.
https://doi.org/10.1038/nphoton.2013.188

[81] Y. Zhao, B. Zhang, B. Yao, Y. Qiu, Z. Peng, Y. Zhang, Y. Alsaid, I. Frenkel, K. Youssef, Q. Pei, X. He, Hierarchically structured stretchable conductive hydrogels for high-performance wearable strain sensors and supercapacitors, Matter. 3 (2020) 1196-1210. https://doi.org/10.1016/j.matt.2020.08.024

[82] X. Fan, W. Nie, H. Tsai, N. Wang, H. Huang, Y. Cheng, R. Wen, L. Ma, F. Yan, Y. Xia, PEDOT:PSS for flexible and stretchable electronics: Modifications, strategies, and applications, Adv. Sci. 6 (2019). https://doi.org/10.1002/advs.201900813

[83] S. Savagatrup, E. Chan, S.M.R. Garcia, A.D. Printz, A. V. Zaretski, T.F. O'Connor, D. Rodriquez, E. Valle, D.J. Lipomi, Plasticization of PEDOT:PSS by common additives for mechanically robust organic solar cells and wearable sensors, Adv. Funct. Mater. 25 (2015) 427-436. https://doi.org/10.1002/adfm.201401758

[84] P. Li, D. Du, L. Guo, Y. Guo, J. Ouyang, Stretchable and conductive polymer films for high-performance electromagnetic interference shielding, J. Mater. Chem. C. 4 (2016) 6525-6532. https://doi.org/10.1039/C6TC01619G

[85] http://refhub.elsevier.com/S0079-6700(22)00046-6/sbref0257.

[86] M. He, Y.J. Lin, C.M. Chiu, W. Yang, B. Zhang, D. Yun, Y. Xie, Z.H. Lin, A flexible photo-thermoelectric nanogenerator based on MoS2/PU photothermal layer for infrared light harvesting, Nano Energy. 49 (2018) 588-595. https://doi.org/10.1016/j.nanoen.2018.04.072

[87] J.P. Jurado, B. Dörling, O.Z. Arteaga, A. Roig, A. Mihi, M.C. Quiles, Solar harvesting: A unique opportunity for organic thermoelectrics?, Adv. Energy Mater. 9 (2019). https://doi.org/10.1002/aenm.201902385

[88] J.P. Jurado, B. Dörling, O.Z. Arteaga, A.R. Goñi, M.C. Quiles, Comparing different geometries for photovoltaic-thermoelectric hybrid devices based on organics, J. Mater. Chem. C. 9 (2021) 2123-2132. https://doi.org/10.1039/D0TC05067A

[89] D.A. Fotouh, B. Dörling, O.Z. Arteaga, X.R. Martínez, A. Gómez, J.S. Reparaz, A. Laromaine, A. Roig, M.C. Quiles, Farming thermoelectric paper, Energy Environ. Sci. 12 (2019) 716-726. https://doi.org/10.1039/C8EE03112F

Keyword Index

About the Editors

Dr. Inamuddin currently holds the position of Assistant Professor at the Department of Applied Chemistry at Aligarh Muslim University in Aligarh, India. His academic journey includes the completion of a Master of Science degree in Organic Chemistry from Chaudhary Charan Singh (CCS) University, Meerut, India, in 2002. Subsequently, he earned Master of Philosophy and Doctor of Philosophy degrees in Applied Chemistry from Aligarh Muslim University (AMU), India, in 2004 and 2007, respectively. With a diverse research background spanning Analytical Chemistry, Materials Chemistry, and Electrochemistry, Dr. Inamuddin has particularly focused on Renewable Energy and Environment. Throughout his career, he has actively contributed to various research projects, securing funding from prestigious entities such as the University Grants Commission (UGC) and the Council of Scientific and Industrial Research (CSIR), both under the Government of India. Recognized for his accomplishments, Dr. Inamuddin received the Fast Track Young Scientist Award from the Department of Science and Technology, India, for his work in bending actuators and artificial muscles. Additionally, he was honored with the Sir Syed Young Researcher of the Year Award in 2020 by Aligarh Muslim University. His prolific scholarly output includes the publication of 215 research articles in reputable international journals and nineteen book chapters in knowledge-based editions by esteemed international publishers. Furthermore, he has edited 180 books with publishers such as Springer (U.K.), Elsevier, Nova Science Publishers, Inc. (U.S.A.), CRC Press Taylor & Francis Asia Pacific, Trans Tech Publications Ltd. (Switzerland), IntechOpen Limited (U.K.), Wiley-Scrivener (U.S.A.), and Materials Research Forum LLC (U.S.A). Dr. Inamuddin actively contributes to the academic community by serving on various editorial boards of journals and holding positions as an associate editor in well-regarded publications. His involvement extends to guest-editing special thematic issues for journals by Elsevier, Bentham Science Publishers, and John Wiley & Sons, Inc. Having participated in and chaired sessions at international and national conferences, Dr. Inamuddin brings a wealth of experience to his role. He has also served as a Postdoctoral Fellow at Hanyang University in South Korea, focusing on renewable energy, and at King Fahd University of Petroleum and Minerals in Saudi Arabia, concentrating on polymer electrolyte membrane fuel cells and computational fluid dynamics. A life member of the Journal of the Indian Chemical Society, his research interests encompass ion exchange materials, sensors for heavy metal ions, biofuel cells, supercapacitors, and bending actuators.

Dr. Tariq Altalhi serves as an Associate Professor in the Department of Chemistry at Taif University, Saudi Arabia. He earned his Ph.D. from the University of Adelaide, Australia, in 2014, receiving the Dean's Commendation for Doctoral Thesis Excellence.

Dr. Altalhi has held leadership roles as the head of the Chemistry Department at Taif University and as the Vice Dean of the Science College. In 2015, Dr. Altalhi achieved recognition when one of his works was nominated for the Green Tech awards in Germany, Europe's largest environmental and business prize, ranking among the top 10 entries. He has also co-edited several scientific books. His research group is actively engaged in fundamental multidisciplinary studies, focusing on nanomaterial synthesis and engineering, as well as the characterization of these materials. Their work extends to various applications, including molecular separation, desalination, membrane systems, drug delivery, and biosensing. Dr. Altalhi has successfully established significant connections with major industries in the Kingdom of Saudi Arabia, contributing to the practical implementation of his research findings.

Dr. Mohammad A. Jafar Mazumder has been serving as a Professor of Chemistry at King Fahd University of Petroleum & Minerals (KFUPM), Saudi Arabia. He has extensive experience in designing, synthesizing, and characterizing various organic compounds, ionic and thermo-responsive polymers for corrosion, water treatment, and biomedical applications. Dr. Jafar Mazumder obtained his B.Sc (Hons.), M.Sc (Chemistry) from Aligarh Muslim University, India, MS (Chemistry) from KFUPM, Saudi Arabia, and Ph.D. in Chemistry (2009) from McMaster University, Canada.

In more than 20 years of academic research, Dr. Jafar Mazumder has had the opportunity to work with several international collaborative research groups and has exposed himself to a broad range of research areas. Dr. Jafar Mazumder secured 8 US patents, published more than 85 articles in peer-reviewed journals, 37 conference proceedings, 9 book chapters, and co-edited 4 books with Springers and Trans Tech publications. He is awarded as a Fellow of the Royal Society of Chemistry and Chartered Chemist, Association of Chemical Profession of Ontario, Canada. Besides, Dr. Jafar Mazumder is a member of the American Chemical Society (ACS), Canadian Society for Chemistry (CSC), Canadian Biomaterial Society (CBS), and a life member of the Bangladesh Chemical Society (BCS). In his academic career, he was awarded numerous national and international scholarships and awards including the prestigious Indian Council for Cultural Relations (ICCR) Scholarship from Govt. of India for undergraduate studies in India, Aligarh Muslim University undergraduate & graduate Gold medal, and certificate of excellence from the Ministry of Human Resource Development, Govt. of India, and MITACS postdoctoral fellowship (Canada) for pursuing postdoctoral research in Chemical and Biomedical Engineering.

Currently, Dr. Jafar Mazumder is actively involved in several ongoing university (KFUPM), government (KACST, NSTIP), and client (Saudi Aramco) funded projects in the capacity of principal and co-investigators. His current research interest includes the design, synthesis, and characterization of various modified monomers and polymers for potential use in the inhibition of mild steel corrosion in oil and gas industries and the preparation of multilayered polyelectrolyte coated membranes for the removal of heavy metals and organic contaminants from aqueous water samples. The long-term scientific goal of Dr. Jafar Mazumder is not merely to make science fun and entertaining for people. It is to engage them with a multidisciplinary scientific mission at a deeper level to create a space through which they can interact with scientific ideas, develop connections between science, engineering, and biology, and thoughts of their own to contribute to society. He feels this goal and engaging personality make him a pleasant person to work with and help inspire his co-workers in any professional setting.

www.ingramcontent.com/pod-product-compliance
Lightning Source LLC
Chambersburg PA
CBHW071655210326
41597CB00017B/2214